PRAISE FOR *NEUROMARK*

‘Katie Hart delivers a clear, practical guide that turns insight into how people think and make decisions into everyday marketing advantage.’

Roger Dooley, author of *Brainfluence* and *Friction*

‘A powerful, poignant and practical guide to the role of human brains in modern marketing. With increasing digital options and benefits available to our profession, it is vital that lessons from neuroscience are also applied to truly understand, enable and support the end decision-maker.’

James Sutton, Strategy and Commercial Director, Chartered Institute of Marketing

‘You know that phrase “it’s not brain surgery”, used to describe things that are hard to understand? Well, somehow Katie Hart makes the brain seem understandable and the practical takeaways feel accessible. A lovely, informative read.’

Joe Glover, Co-Founder, The Marketing Meetup

‘From building understanding of our consumers and their decision-making at an emotional level to empowering us to use this insight to truly build brands and experiences that connect memory structures, this is a great introduction into the world of neuromarketing.’

Abigail Dixon, Founder, Lead Consultant and podcast host, The Whole Marketer

‘Katie Hart is a practising neuromarketer, who speaks regularly on the subject. This is a clear, concise and, above all, readable guide to this fascinating subject.’

Kiran Kapur, CEO, Cambridge Marketing College

Neuromarketing

Practical insights for improving customer engagement

Katie Hart

First published in Great Britain and the United States in 2026 by Kogan Page Limited

Kogan Page
Kogan Page Ltd, 2nd Floor, 45 Gee Street, London EC1V 3RS, United Kingdom
Kogan Page Inc, 8 W 38th Street, Suite 902, New York, NY 10018, USA
www.koganpage.com

EU Representative (GPSR)
eucomply OÜ, Pärnu mnt 139b–14 11317, Tallinn, Estonia
www.eucompliancepartner.com

Kogan Page books are printed on paper from sustainable forests.

ISBNs

Hardback 978 1 3986 2278 4
Paperback 978 1 3986 2277 7
Ebook 978 1 3986 2279 1

British Library Cataloguing-in-Publication Data
A CIP record for this book is available from the British Library.

Library of Congress Cataloging-in-Publication Data
A CIP record for this book is available from the Library of Congress.

Typeset by Integra Software Services, Pondicherry
Printed and bound by CPI Group (UK) Ltd, Croydon CR0 4YY

To…You.

Yes, You. In celebration of your curiosity, your willingness to consider doing things differently and your determination to do the best you can for your organization and your customers.

I applaud you, and am grateful to be a part of your journey.

CONTENTS

PART TWO
Putting theory into practice

LIST OF FIGURES

Introduction

How to use this book

Do not worry, this is not a set of instructions for you to follow – let's face it, I am sure the last thing you need is one more person telling you what to do! Instead of instructions, this is a set of *intentions*. My intentions. For what you will get from this book, and some of the ways you can approach getting them.

This book you now hold in your hands is designed to give you an introduction to the fascinating field of neuromarketing. In order to get the best from it, I have made two assumptions:

1. I have assumed you are already familiar with marketing, and have probably been practising it for a number of years. However, you are curious. You want to keep learning, keep growing and keep improving what you do and how you do it. Somewhere, somehow, you have come across the concept of neuromarketing, and now here you are, ready to discover more.
2. I have assumed you are less familiar with the field of neuroscience. You may have a general understanding of what it is and some areas it is impacting our daily lives, but not in any great depth or detail. What you do know, though, is that surely finding out more about the way your audiences' and prospects' brains work is going to help you understand, deliver and communicate more effectively with them.

If these two assumptions feel like a good reflection of your current situation, then 'Welcome'! You have come to the right place. And, if you feel they don't quite reflect where you are at the moment, you too are still very welcome here. I will do my best to make sure you too get value from what is contained in these pages as, if nothing else, we can all be more enlightened consumers!

Either way, you are here, and for that I am grateful!

The question then becomes, where do you go next?

You see this book is split into two sections. Each one is designed with a specific purpose in mind, and each one has a contribution to make in its own right.

Part 1: Chapters 1 to 6

In the first few chapters of the book, I will introduce you to the world of neuromarketing. I will share with you some of the reasons it is such a powerful force within marketing, and help you to appreciate the ways it can leverage greater results for you (Chapter 1). I will then take you on an exploration of the tools and methods that neuromarketers use, and help you to understand how we have come to learn what we now know (Chapter 2).

Then, we will consider some of the core processes within the brain that have the greatest significance for us within our profession. These include:

Attention (Chapter 3) – why do some things capture our attention and get noticed by us, when other things do not? What do we know about these processes which will help us stand a better chance of getting our materials and content noticed?

Emotions (Chapter 4) – what are emotions, and what role do they play when it comes to your customers? Which areas of the customer journey do they affect, and how do we work with them to achieve better results for our strategies and campaigns?

Memory (Chapter 5) – why is it that we remember some things better than others? Even the things we might try really hard to recall can often evade us. And yet theme tunes from a TV show we watched as a child, we can still be word-perfect on! So what makes the difference, and how does understanding this help us market more effectively?

Decision-making (Chapter 6) – what actually happens within our brain when we make a purchasing decision? What factors do we take into account, and how can they be influenced by things which may be internal and external to us? And what happens to those processes when we are in a B2B decision-making situation?

Part 2: Chapters 7 to 15

In this section of the book, I will help you to explore and discover some of the areas of greatest opportunity for neuromarketers. You will consider some of the key elements that I use and apply for my clients, and understand some of the variations you may want to apply in your own work.

This section is designed to build upon what you have learned about brains and processes in Part 1, and, in some cases, it will directly refer to content you have encountered there. However, if you are not interested in understanding *why* I am making the claims and suggestions I am, then you can just go straight to this part of the book. It will still deliver lots of 'aha' moments, I assure you, but without you needing to get into the background bits so much. The choice is yours...

We will start by taking a fresh approach to getting to know your audiences (Chapter 7). You will understand why this may not have been working very well for you so far, and discover new approaches and methods you can incorporate going forwards.

In Chapter 8, we will explore the power that images have for our prospects and customers. You will see the vital role they play in determining the level of processing which gets applied, and be introduced to new ways of considering and selecting the images you use.

Then we move on to examine colour (Chapter 9). What information does our visual system impart when we experience colour vision, and what does that do to our thoughts and behaviours? The answers are fascinating, and the results may exceed your wildest expectations!

Next, we come to examine the role of words and text (Chapter 10). Copyrighting is a skill many of us have honed over the years, but do you realize the challenge this format creates for your audience? Thankfully, there are some words that appear to be able to help us, and you will examine a few of them in more detail here.

Finishing with the information that we receive through our eyes, we then move into the other senses. Chapter 11 examines the process and opportunities of auditory or sound sensations. What are the different forms of sounds we can utilize, and how can we better understand the specific effects and results they may be likely to deliver?

In Chapter 12, we focus on the sense of touch. What role does this play in the information we receive and the perceptions and decisions we make? Specifically, how does this change now that much of our content and interactions may be taking place online? Is there still a place for tactile approaches alongside our digital strategies?

The final two senses of smell and taste will be covered in Chapter 13. These provide great opportunities for you to intentionally craft and manage the experiences your audience has. You will learn that the benefits of implementing these insights can be vastly profitable, and are relevant to all business models and sectors. So, do yourself a favour and do not skip this one!

Then in Chapter 14, we get real. We stop considering the processes in isolation, and we examine the way our brain approaches the holistic processing of information. What happens if there are conflicts between what our senses are reporting? Can our own brains deceive us or be easily deceived themselves?

Finally, we pull it all together (Chapter 15). I will share with you some considerations, some resources and some suggested approaches to get you started with implementing what you have discovered. I know it can feel overwhelming when you first learn so many great new approaches and ideas, but this chapter will help you to prioritize where to start, and make sure you stay on the right track.

So, there you have it. You can dip into the topics or areas which you feel most drawn to, or you can engage with all the content in a more linear way, going from start to finish. The choice is yours.

Whichever way you choose to access the content, I have one appeal to make to you. Please, please, please, do not just read this book and do nothing with it. The investment you will make in reading it is precious time that you will not get back. So do all that you can to make sure you act on what you learn.

In order to support you with this, I have developed a series of resources for you. There is one for each chapter of the book, and you can access them all online (www.katiehart.co.uk/bookbonuses). Available in both digital and printable versions, the resources will help you to take some of the core elements of content from the chapter, and directly apply them to your role, your organization, your customers.

Some will take longer to complete than others. Some will be easier to complete than others. However, please do not avoid the harder sections. This is often where less committed people will bail out. Please, do not be one of them. You have come this far, you owe it to yourself (and your customers!) to finish the job properly.

As you complete each resource at the end of each chapter, you will start to build your own library of insights which you can refer back to, amend and develop as your knowledge increases. In no time at all, you will be a more confident and competent marketer, as you incorporate neuroscience into all of the decisions and actions you make going forwards. I am so excited for you!

However, now, it is over to you.

Are you ready to get underway? Then let's dive in...

PART ONE

Insights into the customer brain

1

Welcome to neuromarketing: The benefits and opportunities

Neuromarketing is a gamechanger. I appreciate that is a bold claim to make, but I stand by it. Once you start to understand and experience the difference neuromarketing can make, you will never look at marketing the same way again. You will also never approach marketing the same way, as the insights contained within these pages will provide you with a very different framework to consider and apply going forwards.

I hope the fact that you are here and reading this means that you are already excited about what neuromarketing can do for you. Whether you have explored it before, or are only just hearing about it for the first time now, you are equally welcome here. My intention is that you will gain not only background information, but also practical insights that you can begin to apply and test for yourself. Now, before we get underway, I don't want to assume that we have all arrived at this page from the same point. So I will assume nothing, and seek to give you some of the background and examples of the insights that, to me, make neuromarketing such a compelling area to be explored.

First, let's just be clear on what we mean by the term 'neuromarketing'. Simply put, it is bringing neuroscience and the discoveries that have been made within this field into marketing. Now, strictly speaking, neuroscience is defined as the study of the nervous system. So, it is not just the brain that we may be learning about or measuring, it's actually the whole of our nervous system, and that gives us lots of opportunities that we'll come to look at a bit later on. However, for most of what we're going to be doing in this book, we will be focusing on the brain. You see, the more we observe, learn and understand about our brains, the better chance we have to be able to present our marketing messages in engaging, exciting and impactful ways. By bringing neuroscience into marketing, we are using unparalleled insights to give us access to information that many of our audiences don't actually have themselves.

Benefits

Have you ever experienced a situation where you have diligently done the research, engaged with your audience and stakeholders, and delivered precisely what it is they said they want... only to get crickets? Silence? Maybe a few orders or responses trickle in, but the results are a long way from what you were hoping, and predicting, to be the case. Well, you are not alone. And there is a good reason for this. David Ogilvy, often considered one of the 'Fathers of Advertising', famously summarized this challenge by stating that there are clear differences between what customers think, feel, say and do.[1]

The first thing we need to appreciate is that they are not doing this intentionally. They are not being deceitful or trying to manipulate the outcome in any way. It is simply because they do not know the key information that we are trying to reach. They do not know what goes into making the decisions that they do, day in, day out. None of us does. You see 95 per cent of what goes into the decisions we make is actually information that is below our conscious threshold.[2] This means that, at best, if we do really robust research and we spend huge amounts of time and effort and energy with our audiences, at best they can only give us access to 5 per cent of the reasons that they make the purchasing decisions that they do, or share 5 per cent of the motivations behind their behaviours. How do you feel about that figure? I know I wouldn't want to base any business decision or recommendation I make to my clients on just 5 per cent of the information. So, we need to do something different if we are to reduce the amount of wastage within marketing budgets, and provide greater certainty for future outcomes. What we need to do is bring in a very different set of methods which allow us to access and interrogate the 95 per cent that influences and informs the decisions our audiences make. These methods are based in neuroscience.

Opportunities

Now do you agree that there is a very compelling case for including neuromarketing insights into your marketing plans and decisions? Great, so let's get to work then. I think a helpful place to start this exploration is with a question. Why is neuromarketing now a 'thing'?

What I mean is, why are we only now considering the role of the consumer's brain so fundamentally within the marketing decisions that we make? The simple answer, despite how facetious it may sound, is... because we can. Because for the first time, we are now actually able to study the human brain in a really meaningful and informative way.

You see, if you think about it, the brain is a very difficult organ to study. If you could take out a working heart, but leave it connected to its host body, and place it on the table in front of you, what would you see? You would see it pumping, rhythmically. Ba dum... ba dum... ba dum. This would give you suggestions on what the function of the heart is. You could start to see differences between hearts, and learn which ones are more healthy and which are diseased. The same goes for our lungs. If you could take out a working pair of lungs, but leave them connected to their host body, and place them on the table in front of you, what would you see? You would see them expanding and contracting, filling with air and emptying again, rhythmically. Inflate... deflate... inflate... deflate... inflate... deflate. This would give you suggestions about what the lungs do, and again you could study differences between them and learn about healthy and diseased lungs.

Now consider the brain. If you could take out a working brain, but leave it connected to its host body, and place it on the table in front of you, what would you see?

Nothing.

No movement at all. Nothing to provide any indication of what is going on within its bizarre pink and wrinkled structure. So, for many, many years the brain has remained seriously misunderstood by the anatomists of our past. Over the decades, various methods were made to try and interrogate the workings of this mysterious mass, but with slow and unreliable progress. Barbaric corkscrew-like devices were used in a process called trepanning, to remove sections of skull and expose the brain beneath. These areas could then be manipulated to try and stimulate changes within the patient; the specific nature of this varying according to the nature of the symptoms they presented with. Rest assured, these treatments were not particularly common or 'dished out' on a regular basis. No, these would have been reserved for people who were perhaps having visions, fits, seizures or who were suffering from what we would now class as a severe mental illness. However, during these times, the 'treatment' may have been provided as a result of more spiritual concerns for the individual, often including the possibility that they had been possessed.

As you can imagine, such processes did little to advance our understanding of the inner workings of the human brain. But other events did. Sometimes, they were events that were entirely unintentional, accidental even. Step forward Mr Phineas Gage.[3]

Phineas Gage

Mr Gage is often quoted, cited and referred to within neuroscience history, as his situation provided great insights and great steps forward in terms of our understanding of the human brain. But he was not a biologist, a surgeon, a chemist or indeed any form of scientist. Nor was he a religious or spiritual leader in any way. Mr Gage was a construction worker on the American railroads during the mid-1800s. Part of his job involved the removal of sections of rock and cliff face, in order to allow for the railways to be built. Specifically, this meant large holes had to be drilled into the rock face, explosives (gunpowder) would then be placed into the holes along with a charge, next would go a layer of sand to prevent contact with the volatile explosives and then a tamping rod would be used to pack them all in tight. When the time was right the charge would be lit, the explosives would detonate and the rock would be blasted into smaller bits which could then be removed. Simple.

However, on 13 September 1848 something went badly wrong. It is alleged that Phineas omitted to put the vital layer of sand into the hole after the gunpowder. So when he packed it tight by ramming it with his tamping rod, the rod must have caught the side of the rocky hole, a spark was created and the explosives detonated. The resulting explosion launched the tamping rod (which was about 1 metre long, 3 cm in diameter and weighed about 6 kg) back out of the drilled hole, and straight though Mr Gage's head. And I do mean straight through it. The rod went in under his left eye socket, and out through the top of his head, before landing on the ground around 25 metres behind him.

Now, I do not know what you think Mr Gage's chances are at this stage. It is 1848, he is in the middle of nowhere and he has just had a large part of his brain removed by an unsanitary metal rod passing through it at speed. I think we can safely say most of us would think the outlook was not good for him. However, the reason we cite him so much in neuroscience is that, remarkably, he survived this incident. Yes, he lost consciousness for a few moments at the time, but was then able to sit up, stand, walk to a nearby cart and ask the driver to take him to the doctor's lodgings. Imagine that... walking, talking and creating intentions, all within moments of suffering such a catastrophic injury to his brain.

Over the next few days, he drifted in and out of consciousness, probably due to infection and his body going into shock. However, contrary to the predictions of the medics treating him, after about 11 days, he turned a corner and started to recover. Within a few weeks his strength was coming back, and within a few months he was ready to return to work. He retained full use of all his limbs, had no issues with speech at all, he could recognize and identify people, and his memory was unaffected. Miraculous.

Well, almost. You see, despite the amazing recovery that Phineas made, he was never quite the same. He is reported to have become coarse, aggressive, disrespectful, lacking in social skills and basically unpleasant to be around. For a few years, he struggled to hold jobs down and appeared to have issues forming and maintaining friendships. Ultimately, a few years later, having moved back home to live with his mother and sister, he died – 12 years after his accident – from what was believed to be an epileptic seizure.

After his death, his skull and the tamping rod that went through it, were placed on display at the Harvard Medical School, where they remain to this day.

So, what did the famous case of Mr Gage, teach us about our brains? In a word... plenty! We learned about the location of core functions such as personality, social behaviour and emotions as a result of him damaging much of his frontal and temporal lobes. We also learned about brain rehabilitation, and ways that the brain can recover from illness and injury, as it appears some of his missing cognitive functions did return after a period of four years.[4]

However, despite all that we gained as a result of Phineas Gage's tragic accident, and subsequent recovery, none of this so far would have given us many insights to include in our marketing strategies, would it?

These meaningful insights have only been made possible as the result of developments and innovations in technology. Thank goodness we are now able to not only explore brains in very exciting, detailed and revealing ways, but we are able to do all of that using non-invasive methods. No surgery required!

Because now we have broken through the mystery, opened Pandora's box and changed the way we view ourselves for ever. We can now observe a working human brain. We can identify the parts of it that are busy and activated when we present different information or the same information in different formats. We can observe the differences that exist in the structure and performance of brains of different ages and with different health conditions. We can observe it when we ask participants to carry out a variety of routine or complex tasks. We can observe it when we trigger people to feel certain emotions or make specific decisions. And all of these observations can be recorded, measured and analysed in robust scientific ways.

In the next chapter I will outline some of the methods used that allow us unparalleled access to the workings of the human brain, and I will also briefly outline some of the key processes the brain uses, and which we need to understand, in order to be able to apply neuromarketing for the greatest effect. However, before we get into that, I want to do two things. Firstly, I want to share some statistics and insights with you, to allow you to begin to appreciate the fabulous organ we are going to be learning about and appealing to. And secondly, I want to address the ethical questions that often crop up regarding neuromarketing. I am sure you were thinking some of them yourself, so let's get those out in the open and dealt with up front, so you can then continue your neuroscience journey with a clear conscience.

Your brilliant brain

First, those statistics I promised you. I am sharing these with you so you have some appreciation for the magnificent, flawed, complex structure we are going to be working with. Let's begin by just considering what it actually does. Well, for a start, it has just read that sentence for you, which in itself is a very complex undertaking. The physical elements involved in reading and understanding language are phenomenal, before we then start commanding our brain to actually 'think' about something. Then, add into the mix the fact that all this time your brain is also controlling your core body temperature, monitoring your digestive functions, keeping your eyes blinking periodically, and ensuring it receives enough oxygen and nutrients in order to keep itself working. It can recall memories from years ago and make plans for years ahead. It determines our mood, perceptions, motivations and physical actions. It influences the relationships we have, the hobbies and interests we pursue and the environment we choose to locate ourselves within. It can very quickly conjure up images that it may never have seen or put together before... if you don't believe me, just imagine a purple and yellow giraffe, wearing a tiara, dancing on an iceberg, holding a frying pan between its front teeth. See?!

This is all possible due to the incredible complexity that sits between our ears. An average human brain is thought to contain about 86 billion neurons.[5] Eighty-six billion. But wait. Each of these neurons has the potential to connect to thousands of other neurons.[6] That is the equivalent to standing on the stage at the iconic Peacock Theatre in London and being able to physically connect to every single member of the audience. Only, each of them can also connect to thousands of other people. Can you see how this soon creates an incredible network of connections and possibilities?

Our brains are also phenomenal processors. In fact, they process approximately 11 million pieces of information per second... and that is just from our senses. But, we are only likely to ever be consciously aware of between 40 and 60 of those.[7] That means, there are at least 10,999,940 bits of information being processed in our brain that we know nothing about. Every. Waking. Second. This creates the first truth that we need to acknowledge – the sheer volume of information and activity that takes place within our brain that we consciously know nothing about.

This can be hard for us to accept. I think most of us like to believe that we are consciously in control of all that we decide and do. But the harsh truth is that we are not. The brain has developed very efficient, and usually effective, processes that allow it to operate below the radar of our conscious awareness. You see our brain is a very expensive resource for our body to run. Although it weighs an average of 2 per cent of an adult's total body weight (3 lbs or 1.35 kg), it typically accounts for more like 20 per cent of all the resources that a body consumes.[8] So, on average, 20 per cent of the oxygen we inhale, 20 per cent of the calories, all to fuel an organ that is only 2 per cent of our body weight. Some of the most 'expensive' functions within the brain are those we apply conscious awareness to, so by limiting these, the brain is able to operate as efficiently as possible.

A little experiment

Let me give you an example. If I ask you now to, just for a moment, put this book down and cross your arms please. Go on, do it. Now, please do it again for me as quickly as you can – imagine that you are a stroppy child who has not got their way, and see how fast you can aggressively fold your arms this time. Were you quick?

Now, I want you to do the same again, as quickly as you can, but this time I want you to fold your arms the opposite way. Go...

Have you done it yet? How did it compare to the previous direction? Usually it is much slower and more 'clunky'. I bet it also took a lot more mental effort to work out how you usually fold your arms, and therefore what you need to do differently to fold them the other way. This is because our brain uses things called heuristics. They are essentially short cuts that the brain develops and applies to conserve its precious resources, whilst still allowing us to get on with our day. Can you imagine how slow we would be if we had to apply the same conscious effort you needed to fold your arms the 'wrong way' to picking up a drink and taking a swig, or talking to someone whilst walking alongside them? Our days would be impossible. Most of us could probably never leave the house. So, the brain uses heuristics to save us the mental capacity. We form habits, so we don't need to think about what we are doing. We just repeat things until they become automatic for us – dry our body in the same order when we get out of the shower, put the same leg into our clothes first each time, call in at the same coffee shop on our way into the office.

For marketers, these heuristics are both a blessing and a curse. You see, many of these heuristics will be affecting the decisions we make on a daily basis. Or, to be more precise, the lack of decision that we make. You see if I usually purchase brand X of teabags, this may be becoming a heuristic, and so I will apply little thought to the action when I next need to buy some. So brand X will benefit from this action, and the repetition that is building that as a heuristic. But if brand Y wants me to buy their products instead, they are going to have to work hard to divert me away from this 'easy' course of action, and apply the conscious effort involved in doing something different.

The two parts of the brain

Many of these insights and opportunities have come about as a result of understanding that our brain should really be viewed as two different parts. No, I am not talking about the different hemispheres here, I am talking about the different layers that exist within the brain. The neo-cortex ('neo' meaning new) is the bit we might usually picture when we think of a brain – it is the pink, wrinkled structure that is indeed divided into two hemispheres. However, underneath that is a much older part of our brain, which can be traced back millions of years into our evolutionary past. The neo-cortex is more recent in terms of our development, and operates very differently from its ancient counterpart. Let's just look at some of those key differences now.

The neo-cortex is considered to be very smart. It can do phenomenal things like work out what 35 × 17 is without using a calculator... I promise you, it can! It can create intentions for our future, recall past experiences, process language and even consider its own existence. However, it does take effort to do some of these. Did you try to work out what 35 × 17 is? If not, try it now and see how much of your precious cognitive resource you need to attempt it. It's OK, I will wait...

Have you done that? Really, have you? If so, you may have felt how involved that process was. And all the time you are trying to complete it, you were not likely to be thinking about anything else like what to have for dinner or whether you have received a response to 'that' email yet or not. Most of your capacity was taken up with focusing on the task. So, to a certain extent, we can see that we are able to consciously control this part of our brain. As a result of this, it is not always 'on', i.e. when we go to sleep, this part becomes much less active. (And if you need completion before we can move on, the answer is 595.)

In contrast to this, we have the older part of our brain that you may hear referred to as the reptilian brain or the limbic system. This part is below our conscious control, and it operates at a much faster speed than our neo-cortex. This is because it is essentially focused on ensuring our ongoing survival. So, although it is fast, it is actually quite limited. However, it is also permanently 'on'. If you are asleep in bed at night and there is a 'thud' downstairs, it is this part of your brain that will rouse you and bring you back to a conscious and reasonably alert state. Probably with your heart already racing and your senses heightened in preparation to deal with whatever caused it. That is what this part of our brain does.

So, we have two parts that operate at very different levels and speeds. This is fundamental to understanding the opportunity for neuromarketing. Instead of researching and communicating with the neo-cortex, we have to move our focus to the older, subconscious part of our audiences' brains. This is the area that will determine whether our marketing materials get noticed and remembered, whether there is sufficient desire to motivate us to act and ultimately whether we decide in favour of purchasing or not. This is the area we will be learning to market to in the coming pages, and finding out more about the ways we can understand the factors that influence the processes it adopts.

One area that I find people struggle most with this realization is when we are dealing with business-to-business (B2B) marketing. We believe that in B2B scenarios, people make decisions differently. The process of forming a decision making unit (DMU) or employing professionals to conduct our procurements gives us the illusion that we are making 'better' decisions, and are less likely to make mistakes. However, the subconscious parts of our brains that inform the decisions we make are not concerned about whether we are at work or at home. They utilize the same approaches and are influenced by the same factors wherever they are. So, although neuromarketing has significantly changed the way many B2C organizations approach their marketing, there is still a lot of work to be done within the B2B sector. The good news though is that as a result of this, there is a lot to be gained for those organizations and individuals who embrace it.

How ethical is all this?

So, now on to the ethics of neuromarketing. It is only right that we address this up front, as I am sure we all want to conduct our marketing ethically and within the parameters of decency. I also think this is important to cover, as we are all consumers too – we need to know that we are not being manipulated or cajoled, just the same as every consumer has a right to expect the same. So, is neuromarketing ethical?

The short answer is yes. However, I am sure you will not just take my word for it, so let me outline the reasons why I confidently state that. First, all marketing is essentially about trying to persuade people to do something. That may be making a purchase, signing up for a newsletter, attending an event, sharing details of my preferences or behaviours, joining a cause – you name it, we marketers want people to do it. And neuromarketers are no different. We are still in the same profession. However, we have some additional tools that we can bring to the fore. Rest assured though, none of these tools will persuade someone to buy something they had no intention or interest in previously. That fear is unfounded.

Instead, what we *are* able to do is learn how to support people in their decision-making processes, facilitate their engagements with an organization, reduce the volume of materials that are frustrating and irrelevant to them, and ultimately help them to feel seen and appreciated. I firmly believe that consumers deserve better. We deserve to be treated as human beings and engaged with in interesting and meaningful ways. No more should things be done *to* us, but they should be done *for* us. And this is where neuromarketing provides unparalleled opportunities. So read on to find out more.

Chapter summary

1. Neuroscience allows us to understand our customers at a new, much deeper, level than we have been able to access and interrogate before.
2. Some 95% of what goes into all of the decisions we make is below our conscious threshold, so no amount of interviews, focus groups or surveys will ever enable us to access this information from our audiences.
3. We are now able to study and observe working human brains, in reliable, scientific and non-invasive ways.
4. The human brain is a fabulously complex structure, which processes approximately 11 million pieces of information from our senses each second.

5 Although it weighs roughly 2% of our body weight, our brain consumes more like 20% of our bodies' resources.

6 Brains use short-cuts called heuristics to help themselves operate at a very efficient level.

7 Our brain is made up of a neo-cortex, which is very clever, but slow, and our reptilian brain, which is very fast, but limited.

8 Neuromarketing is ethical, and is equally applicable in both B2C and B2B situations.

Notes

1 Ogilvy, D (1963) *Confessions of an Advertising Man*, Atheneum, New York

2 Zaltman, G (2003) *How Customers Think: Essential insights into the mind of the markets*, Harvard Business School Press, Boston

3 Gearhart, S (2025) Phineas Gage: American railroad foreman (25 June) Encyclopedia Britannica, www.britannica.com/biography/Phineas-Gage (archived at https://perma.cc/WD2U-7NUY)

4 Teres, R V (2020) Phineas Gage's great legacy, *Dementia & Neuropsychologia*, 14 (4), https://doi.org/10.1590/1980-57642020dn14-040013 (archived at https://perma.cc/HPP6-6P3D)

5 Azevedo, F A et al (2009) Equal numbers of neuronal and nonneuronal cells make the human brain an isometrically scaled-up primate brain, *Journal of Comparative Neurology*, April 10, 513 (5), 532–41, DOI:10.1002/cne.21974 (archived at https://perma.cc/NSZ4-6UPX)

6 Caruso, C (2023) A new field of neuroscience aims to map connections in the brain (19 January), Harvard Medical School, https://hms.harvard.edu/news/new-field-neuroscience-aims-map-connections-brain (archived at https://perma.cc/FS3R-NU2L)

7 Markowsky, G (2025) Information theory (19 July) Encyclopedia Britannica, www.britannica.com/science/information-theory/Physiology (archived at https://perma.cc/2PMU-4WJB)

8 Padamsey, Z and Rochefort, N L (2023) Paying the brain's energy bill, *Current Opinion in Neurobiology*, 78, https://doi.org/10.1016/j.conb.2022.102668 (archived at https://perma.cc/HNA9-95WJ)

2

Tools of the trade

Context

Having established the case for neuromarketing – what it is and why it provides such unparalleled insights – before we launch in to discovering those insights and putting them into practice, there are two further steps I think we need to take. First, I want you to understand the methods that have been used to discover what we have learned, and, second, we need to consider some key chemical processes within the brain that will crop up time and again as we go through the content.

As a result of the developments outlined in the previous chapter, we can now ask questions that we have never been able to answer before – what goes on inside somebody's brain when they are processing information, engaging with packaging, observing a film trailer, walking past an advert on the London Underground, scrolling their social media feed, listening to a podcast or trying to navigate around a website? All of this is available to us for the first time, meaning we now have access to that vital 95 per cent of subconscious information that informs our decisions and determines our behaviours (see Chapter 1). This is the critical 95 per cent that we ourselves do not have access to remember, so no amount of interviews, surveys or focus groups will enable us to ever impart it to others. How can we, when we don't know it ourselves? But now it is available, measurable and able to be used to give accurate predictions.

Neuromarketing methods

Let's begin by examining the methods neuromarketers use to allow us to observe, record, monitor and analyse these internal processes. However, before we do, I want to be very clear with you about one thing. I am not presenting these to you as a way of encouraging you to go out and start conducting your own neuromarketing research projects. They are often beyond the scope of what most budgets can afford. No, the reason I am taking the time at this stage to explain these is in the hope that as we go through some of the content in the rest of this book or the resources that accompany it, and as you (hopefully!) continue to explore neuromarketing after you have completed reading it, you can form educated and objective opinions about the research you will encounter. You see, all neuromarketing methods offer different advantages and challenges, and it is only by knowing each of these that you can truly judge the merits of some of the many studies you will encounter. So, shall we begin?

Facial analysis (and body language)

Sometimes when we do neuromarketing research, what we are actually doing is using existing research methods and techniques, but instead of just listening to what people say, we are much more focused on how they say it. So, one of the first things we can do is look at our participants' facial expressions and body language. There is a whole fascinating science around facial coding and body language, which you may be aware of to some extent. It exists because when we receive some information through our senses, our reptilian brain processes it and starts to produce a reaction, before the slower parts of our neo-cortex really begin to engage. Therefore, our brain is creating responses that can 'leak' out in the form of facial expressions and body language, before we then put the social brakes on and adapt our response according to what we think might be socially acceptable or appropriate in any given setting. Think about young children. They have not yet developed this social brake in the prefrontal lobe of their brains, so if they feel something, they show it. If they don't like the taste of something, they spit it out. If they don't like what you are telling them, they burst into tears and throw themselves on the floor... even if you are in the middle of a busy supermarket. As we get older, we become more aware of the need to rein in our responses, and so our behaviours usually become more normalized and less extreme. However, this brake takes time to engage, and in those few precious moments before it does, our face and bodies can display huge amounts of information if you know what you are looking for. So, if we are doing research, we might want to video our interviews and slow them right down and start to study and interpret what we can learn, by capturing these raw expressions as they are fleetingly made available to us.

Voice analysis

Similarly, if we are recording interviews that we conduct with actual and potential customers, we might want to examine the audio content in just as much detail as the video footage. By focusing our attention on *how* they say their answers, we can pick up significant pieces of information which again indicate their true attitudes and feelings. It might be a slight pause or hesitation, which is unusual in their regular speech patterns. It might be that they are *not* hesitating, they are actually going the other way and speeding their pace up. It might be that they have slight changes in the tonality or the rhythm of the way they speak. All of these can again give us hints and insights into what their brain might actually be interpreting from the questions, and how they may be feeling about the responses they are providing.

Electrodermal Activity (EDA)

Next, we start to get into the more physiological measures that are conventionally associated with neuromarketing research. One of the easiest ones to use is Electrodermal Activity (EDA), sometimes referred to by the term Galvanic Skin Response (GSR). These use small sensors that are put onto people's fingers, which will give an indication of how that individual is responding to what they see and hear around them. This is because when our reptilian brains process information and trigger corresponding responses, many of these changes can be detected externally in our physiology. For example, if our brain anticipates something is stressful or negative, it will release cortisol in our brain that will trigger physiological changes incredibly quickly. These changes, such as the 'fight or flight' response that I am sure you are already aware of, create responses that can be picked up by measuring changes in the conductivity on the surface of our skin. Essentially, we start to sweat a little, and that addition of small amounts of water on the surface of our skin means it conducts electrical signals more efficiently, and therefore faster. So, by measuring these changes, we can tell if the brain has prompted the physiological changes to take place. This means that, very quickly, we can get a sense of whether people feel positive or negative towards what they're seeing. And don't forget, these are physiological measures that none of us can easily control, so we cannot skew this, we cannot quickly influence this in ourselves, which gives us physiological access to that raw, unconscious response that is so elusive and yet so vital for us to understand.

Eye-tracking

Another common technique you may come across is the use of eye-tracking. This allows us to determine, with varying degrees of accuracy, the specific part of a participant's visual field that they are looking at, and how long they remain focused on it. I say 'with varying degrees' because there are a few different methods used to achieve this. At low levels, use can be made of the cameras that are built into your laptop or mobile device. In these cases, software can use that built-in lens to record and process which parts of the laptop or mobile screen the participant is looking at, and the duration of their gaze. However, the most reliable and flexible options are to use eye-tracking glasses. These have two sets of camera lenses. The first set looks out, like camera lenses that are recording the visual field that the participant is experiencing, so what's in front of them at any given time. But they also have a second set of lenses that are recording the participant's eye, and in particular their fovea, which is the location of the detailed, focus point on the retina. From correlating the two of these together, we get a very accurate image of precisely where in their visual field the participant is focusing. So, if you can imagine being in a supermarket or driving down a city street, or visiting an industry exhibition, using eye-tracking glasses will enable us to identify which of the many, many elements that make up our visual field in this complex environment are actually able to capture and hold our attention. What, of all that myriad of stimulation, is the participant actually drawn to, what do they notice and what do they pay attention to?

Electroencephalography (EEG)

Now, we can start to look directly at the actual functions and activities that are happening within the brain. An electroencephalography (EEG) headset records the electrical activity that takes place within the brain. We have already learned that the brain is made up of billions of neurons, which are surrounded by electrical charge. When the brain sends messages, activity within and between these neurons creates tiny electrical impulses which the EEG headset can pick up. So, the headset is just a way of allowing us to view what the brain is naturally already doing, and the processes it is naturally already using. What is different though is that instead of the early EEG headsets – which were like swimming hats covered in wires, required the wearer to sit very still, and also needed them to be attached to a machine via the said bundle of wires – they are now much more portable. In fact, so much so that we can put one of these devices onto a person in almost any environment. I have studied people at their desks, in board rooms, at networking events and even in exhibition halls using this technology, which gives us many more research options and possibilities.

Functional Magnetic Resonance Imaging (fMRI)

Next, we upgrade to some even more sophisticated technology, which allows us to identify the specific parts of the brain that are working at any given moment – Functional Magnetic Resonance Imaging machines, or fMRI. You may have come across MRI scanners in hospitals where they are used to look at the structure of our bodies. Well, these are *functional* MRI scanners, which enable us to watch the brain while it is at work. We can present information to give people decisions to make, to simulate a variety of scenarios or just to check out different logo visuals, and all the time we can be watching the way their brain is responding. The down side of these, though, is that they are far from mobile, and very expensive to hire. So you need to have deep pockets and very willing/enthusiastic participants if you plan to use this in your proposed research.

Positron Emission Tomography (PET)

Finally, we reach the zenith of neuroscience research, which is often out of reach of even the largest neuromarketing budgets. However, academic discoveries from using this have been insightful, so I will cover it briefly here too. Positron Emission Tomography (PET) is a method of studying the brain that is believed to offer greater levels of accuracy than fMRI scans. However, in order to use PET, the participant needs to be injected with a short-lived radioactive chemical, which can then be observed as it makes its way through the brain. This is great for the depth of insights and detail that can be gleaned, but the challenges usually outweigh the benefits for most of us. You see, in addition to requiring the use of majorly expensive pieces of immovable equipment, this process also needs to be administered by highly skilled medical professionals. Oh, and as soon as you mention injecting people with radioactive chemicals, it is also going to be much, *much* harder to secure participants. Believe me!

Finding the right method

Through utilizing these different technologies and different approaches, sometimes in isolation, and sometimes combined, we have a chance in neuromarketing to really interrogate what's going on in the brain.

However, not all of these measure and record the same aspects, so we are always trying to find the optimum balance and the optimum compromise, for the particular piece of research we might be doing. For example, let's compare EEG and fMRI technologies.

The EEG headsets that I use provide me with 256 readings per second from each of the 14 sensors. If we do a quick sum, that is 3,584 readings per second, and 215,040 readings per minute. From each participant! So, yes, I get huge amounts of data, but I also get very high levels of what we call temporal resolution. That basically refers to the speed at which activity within the brain will be picked up by the sensors. EEG is very quick. It picks up fast and sometimes fleeting responses within the brain. However, because of the way the sensors are located on the scalp it's not so great at giving me reliable information about precisely where within the brain the responses are being activated. If you can imagine, those sensors are recording electrical impulses through the different layers of the brain, so it is only prudent to expect some interference and loss of accuracy, particularly towards the very centre. For this reason, many of the algorithms that convert raw EEG data into metrics that are useful for marketers rely on having correlated tens of thousands of responses from EEG with those gained by fMRI. That way, we can be much more confident about the responses received within each different level of the brain. So, why don't we just use fMRI? Well, the fMRI scanners give us that really accurate spatial resolution, so we can identify in great detail the specific parts of the brain that are actually being activated and engaged. However, the way an fMRI scanner works means there is a delay of about three or four seconds between information being presented to the participant and us being able to see activation on the corresponding brain scans. We therefore need to always try to find the best balance and the best compromise when we're conducting research, to say nothing of the need to operate within sometimes very challenging budgets.

Of course, sometimes that balance is best achieved by using combinations of these methods. For example, if we use eye-tracking glasses and EDA, we are able to not only see which precise areas a participant is looking at and focusing on, but we can also measure how positively they are responding to what they see in that place.

Although in isolation, none of these methods may be considered as perfect, what is really important to remember here is that consistently, across all of these techniques, we are bridging a huge gap. That is the gap that exists between what we might traditionally pick up through using conventional market research methods and what we can pick up physiologically using any of these techniques.

One of the best examples of this is what has now become known as the Iowa Gambling Task.[1] In this task, participants were brought in and given a simple task to complete. They were given a fictitious amount of money ($2,000) and were told that the object of the game was for them to try to have as much of this fictitious money as possible at the end of it. They had four piles of cards, face down in front of them, and they were instructed to turn over 100 cards from any of these four piles in any order that they wanted to. Now, each of the cards, when they turned them over, had an amount of money written on it which was either positive or negative. If it was positive, that amount would be added to their fictitious bank account and if it was negative it would be taken away from their fictitious bank account. So, the participants were given all the instructions and told to start turning over cards.

The piles were of course rigged, so that two of them would deliver a net gain over time, whilst the other two would deliver a net loss over time. Now, while the participants were doing this experiment, they were wearing EDA sensors that picked up on the electrical conductance on the surface of their skin. Remember that these are a good indication of when our brain has detected something negative, which may be construed as being stressful or threatening. So, the researchers now had two different sets of information coming in – the physiological responses and the overtly reported responses. What became apparent is after having turned over in the region of 40–50 cards, participants could overtly report a 'hunch' they had that some of the piles were preferable over others. So, maybe we want to be proud of them as that is quite good going. However, if we look at the physiological information, we get a very interesting response. Here, clear indications of negative or stress responses can be observed on the surface of the skin much earlier. That is to say that when a participant reached out to turn a card over from one of the 'bad' piles, their body was already registering it as being stressful – their body already knew it was going to be bad. And what I think is truly remarkable about this is that these responses started to occur after turning over an average of 11 cards. So, after just 11 cards we at some unconscious level 'know' that the outcome is going to be bad, but we cannot consciously articulate that for another 30 or so cards. That is what I mean when I say there is a gap. There are things we 'know' and perceive unconsciously that researchers can pick up at a physiological level, that we cannot articulate ourselves until much, much later... if at all.

This is where neuromarketing comes into its own.

Remember, 95 per cent of what goes into our decisions is this unconscious information, so anything that we are able to do that gives us access to that inaccessible information is still a huge leap forwards from where we have historically always been. Neuromarketing may not have all the answers and perfect solutions yet, but we are much better off than we were in 2008 when I started working in this field. Give us a few more years and who knows what will be available to us.

Neurotransmitters and hormones

Having examined the methods used to learn about our brains, I now want to briefly cover some of the core chemical processes that we also need to understand if we are to make our marketing more scientific and effective: neurotransmitters and hormones.

I know they may sound imposing, but neurotransmitters are precisely what they say they are. They are chemicals that 'transmit' between 'neurons'; they pass messages between neurons (nerve cells) and other neurons or relevant cells. These chemical messages convey information to and from our brain, to provide information about our environment, regulate physiological processes such as sleep and appetite, and they also affect our mood and motivations. In short, they are vital. Some of the main ones for us to understand, are as follows.

Dopamine

Dopamine is the neurotransmitter that is often associated with our motivation. It is involved in our processes of attention, learning and memory, but you usually just hear it being referred to as part of 'the reward system' within our brain. So, what does that actually mean and how does it affect us?

We have many neural pathways within the brain. These are essentially a series of connected neurons, which relay signals from one area of the brain to another. When we are born, these pathways may not be very defined, but as we repeat actions and receive responses, they become stronger. So, as a baby we learn that when we cry we get attention from our care giver. Each time we repeat this process, the pathway become stronger and, as a result, the signals are communicated between neurons faster. It is much like a pathway across a field or patch of grass. The more it is used, the more defined it becomes and the faster it is to cross it.

Dopamine is part of a number of neural pathways, but the mesolimbic pathway is the one primarily involved in reward and anticipation. So, dopamine gets released when we experience something pleasurable, and it makes us feel satisfied – we get a 'dopamine rush'. However, this state is only temporary, so our dopamine levels soon drop and we return to our pre-pleasure state. We naturally want to feel good again though, and so we seek out the next dopamine hit by repeating the behaviour that made us feel good last time. That could be eating fatty foods, scrolling on social media, going shopping or engaging in more potentially destructive behaviours such as gambling or drug abuse. Can you see how critical the release of dopamine is in driving addictive behaviours?

The key element for us to understand as neuromarketers is that we now know that dopamine is not just released when we receive the reward, i.e. when we eat the food or wear the item of clothing. Actually, the highest spike of dopamine is experienced prior to us receiving the reward.[2] It is the anticipation of receiving it that makes us feel so good. And can you see how useful this would have been to us throughout our evolutionary past? For it is this anticipation that motivates us to put in the effort, do the work and take the steps required in order to receive the reward.

Now there is a concept that we marketers can definitely work with!

Serotonin

Similar to dopamine, serotonin is involved in the regulation of physiological processes such as appetite and sleep. However, we normally hear it referred to in relation to our happiness and feelings of wellbeing. Low levels of serotonin are believed to contribute to feelings of depression and anxiety, whereas high levels of serotonin can cause equally undesirable symptoms such as sweating, shivering and even seizures. However, serotonin is also known to play a vital role in controlling our impulses, so low levels of it are associated with us being more impulsive in our actions.

One of the key differences between dopamine and serotonin is that dopamine tends to affect our mood in the short term whereas serotonin is more about regulating our emotions on a longer-term basis. So we are not talking about short spikes of serotonin as we saw with dopamine; they are more enduring and create longer-lasting effects within our brain and body.

Finally, serotonin is also regarded as being a very social neurotransmitter. It makes us feel confident and good about ourselves, and influences many of our social behaviours from a very early age.[3] These aspects combine to make it an interesting consideration for us within neuromarketing, as such influences can have highly significant effects on our buying decisions.

Alongside these key neurotransmitters, it is also useful for neuromarketers to have a basic appreciation of the following two hormones.

Cortisol

I am sure you may have already heard of this one, as cortisol is a hormone that affects much of our everyday lives – sadly. Cortisol is involved in our stress response, and regulates our immune system and blood sugar levels among other functions. However, it is being referenced here as it also has a role to play in many of our higher executive functions, such as how we make decisions. The more cortisol we have, i.e. the more stressed we are, the more impaired our decision-making becomes. Yes, you read that right. When we are stressed, our ability to make good decisions becomes compromised. Moreover, we trust people less when we are stressed, so we clearly need to do all that we can to minimize the occurrence of this within the brains and systems of our audiences.

Oxytocin

Often referred to as the 'bonding' hormone, oxytocin is naturally released immediately after childbirth, when mothers breastfeed, for example. We can conclude from this that it is a very intensive connecting hormone, building trust, empathy and even generosity in its host. Can you see how this is going to be useful to us? If we want people to feel connected to our brand or our cause, if we want them to trust us and the statements we make, and if we want them to give generously, or even just give us the benefit of the doubt, oxytocin is going to be vital. As more brands are developing their own personalities, the opportunities for harnessing the power of oxytocin is getting greater and more compelling. Trust me.

So, now that you are more familiar with the different methods available to researchers, and the ways they work, let's start to discover some of the insights these methods have revealed. To me it seems logical to begin with the idea of attention, because none of our strategies, messages and offers are going to be successful if we are unable to get them noticed by our audience.

Chapter summary

1 Recording conventional market research can enable us to use facial, body language and voice analysis to gain deeper insights than just what the participant is saying.

2 Different methods used in neuromarketing provide different benefits and challenges for the researcher.

3 Although none of the methods is yet perfect, all advance our understanding of our customer base significantly from what we would previously have been able to capture.

4 There are clear gaps between what we 'know' subconsciously and what we are aware of (and therefore able to accurately report) consciously.

5 Dopamine (the neurotransmitter associated with pleasure and rewards) release is higher when we anticipate a reward, than when we actually receive it.

6 Serotonin is a neurotransmitter that plays a major role in our emotions, the control of our impulses and how we behave socially.

7 Cortisol (a hormone released when we are stressed) levels negatively affect many of our higher executive functions, such as decision-making.

8 Oxytocin (the 'bonding' hormone) helps us to trust people, feel generous towards them and build connections with them.

Notes

1 Bechara, A et al (1997) Deciding advantageously before knowing the advantageous strategy, *Science* 275, 1293–95, DOI: 10.1126/science.275.5304.1293 (archived at https://perma.cc/FYS4-TKXT)

2 Sapolsky, R M (2017) *Behave: The biology of humans at our best and worst*, Penguin Press, New York

3 Kiser, D, Steemers B, Branchi, I and Homberg, J R (2012) The reciprocal interaction between serotonin and social behaviour, *Neuroscience & Biobehavioral Reviews*, 36 (2), 786–98, DOI: 10.1016/J.NEUBIOREV.2011.12.009 (archived at https://perma.cc/2DC8-AM7M)

3

Attention

Context

You now know how busy your brain is each moment of every day. So how, in this complicated and competitive environment we inhabit, do some aspects cut through the noise and capture our attention? What influences that filter, and what can we do to improve the chances of our marketing materials making the cut?

I am sure you have heard the phrase 'paying attention'. Maybe it takes you back to your school days when someone would stand at the front of the class and instruct everyone to 'pay attention Year 5' (please tell me that wasn't just me!). Well, this phrase is surprisingly appropriate, because attention is a very expensive commodity for all of us. As we have already learned, the brain is an expensive organ for our bodies to run, its consumption of our incoming resources is totally out of proportion to the size of its structure – 10 times more in fact. And if we study the brain when it is active and engaged, we can see that many of these precious resources are used in processes such as the conscious application of attention.

So, what do we actually mean by 'attention'? What is it we are actually referring to when we talk about 'attention' within our brain? Well, strictly speaking, there is not just one thing we are referring to, as there is not just one type of attention – according to psychological research, there are actually several.

Types of attention

Sustained attention

Let us start by exploring perhaps what most of us think of when we use the word 'attention'. It is what psychologists refer to as 'sustained attention'. This is the form it takes when we are completely immersed, when we are concentrating, when we are focusing on just one thing, and so we exclude a lot of what else is going on around us for a sustained period of time. Maybe you are concentrating on reading an important document, focused on watching your child's performance in a sports fixture or engrossed in a very tense part of a film. Sustained attention can therefore be seen to be very immersive – when we are in this state, we are *really* in it. We prioritize one aspect of our sensory world, and divert all of the precious cognitive resources we've got, into that task. Consciously.

Alternating attention

Next, we need to consider a phenomenon that many of us experience on a very regular basis. Psychologists term this one 'alternating attention'. This is where we switch, or alternate, our attention between two tasks, which often require two different cognitive skills. So we might be drafting some content and then we break off from that to read a notification that has just popped up, before we then return back to the content we were writing. The ability to quickly and effectively return to the task we were originally doing is a defining factor of alternating attention. Personally, I have found that my skills in this area have increased drastically since becoming a parent! The ability to stop what I was doing, change to doing something new and then being allowed to return to my previous task is tested on a daily, if not hourly, basis now. Interestingly too, if we look at the way activity happens within the brain, quite often when we think we are multitasking, we are actually switching. We are busy doing one activity, then we switch to accommodate the second task, before quickly switching our attention back to our original activity.

Divided attention

Sometimes, we can actually divide our attention and efficiently carry out two (or more) tasks simultaneously. Maybe we are listening to a podcast whilst we are working out, or holding a conversation with a passenger while we are driving a car. This form of attention is not easy for our brains, though, particularly if the attention that we are asking our brain to apply to something requires the same part of our brain to perform two different things. For example, do you sometimes struggle to write down an important point in a meeting, whilst you are also trying to respond to it verbally? The reason this causes such trouble for us is because we are actually trying to use the language processing part of our brain to do two things – write what has been said and also construct our response to it. What actually happens is we reduce our ability to do either thing efficiently, so both our spoken sentence and our written notes become compromised and slower. Divided attention works best for us when the tasks do not require the same regions of our brain to be involved.

Selective attention

Finally, we have what psychologists refer to as 'selective attention'. This really is where we are going to be concentrating as marketers, because selective attention is what captures our attention in any given moment. Of all the stimuli that exist around us, we select only a few to give our expensive attentional resource to, and accept that the other distractions will get ignored. And that is why it is interesting for us within this field. We want to understand the processes for deciding what it is that will receive our attention – how does this selection get made, and what can we do to improve our chances of reaching that threshold?

Before we get into the details of these questions, I want to remind you about something we discussed in the first chapter, because it is very relevant here. That is the concept of heuristics. Do you remember these, the shortcuts the brain uses to work efficiently? They really come into their own here because, as we have discovered, paying attention to things demands a lot of brain activity. So, in order to enable it to work more efficiently, our brain uses heuristics, which in turn means we need less attention applying in order to complete everyday tasks. When we are carrying out a task for the first time, such as learning to drive a car or play a musical instrument, the process takes huge amounts of our attention. However, the more we rehearse them, the more familiar they get, and the more automated they become. Until before you know it, you can not only drive the car, but you can have music on or hold a conversation at the same time. All because the required level of processing is vastly reduced, and our brain is using its own finite resources, as efficiently as it can.

The defining aspect of attention I want to discuss with you is one I am sure you are already aware of. It is the idea that we can have variability in terms of the intensity of attention we consciously apply. We effectively have a spectrum, along which we can choose to apply a little bit of attention to something or, at the other end, we can go all in and apply all of our attention to it. Now, I feel like I have got to know you well enough to share something personal with you here. I'm sure you would never do this, but I know I've had conversations with people over the years where, between you and me, I may have only been paying them a light level of attention. You know the kind of conversations I mean? No, of course you don't, I am sure you would never be so rude… or tired or hungry or bored or stressed or… (you get the picture!). Let's go the other way then. Have you had conversations with people who literally capture your attention and you are fixated, hanging onto their every word. Maybe they have been a speaker at a conference, or just a very charismatic and enthusiastic colleague. Or what about if you are watching the latest episode of *The Traitors* – there may be parts of that which play in the background whilst you check your phone, but in other parts, you are totally attentive as you don't want to miss anything that gets said. Either way, if you have experienced it, you know what I mean when I say that we need to acknowledge that there are degrees of attention that we can apply.

When we look at models of attention, and cognitive psychology has studied attention for decades, some of the things that are interesting are the ways they describe attention. All the models of attention relate to the fact that we are filtering, on a massive scale. There is a process that screens out what we pay attention to. So you can sometimes hear it called a bottleneck or a lighthouse or even a torch effect. Essentially, if you think of it as a lighthouse analogy, it's realizing there is a whole environment around us but, at any given point, the beam of light is fixated on one small part of that. So one aspect of what is available to us will hold our conscious attention, and the rest of it is dialled down. The rest of it we aren't consciously aware of. It's not to say we're not unconsciously aware of it, but consciously we won't be aware of those aspects.

How does attention work?

I am sure by now you are starting to ask yourself (or more accurately me!) how these processes actually work. Let's begin by reminding ourselves that we have two levels of activity to consider here. The first is the level we *can* consciously influence and direct. We can choose to really tune in and listen to the person leading the meeting, or listen to what our partner is telling us about their day, or listen to the headline story in the news. However, all the time we are choosing to attend to that, we are also choosing *not* to attend to a huge variety of other aspects that are happening around us. Remember those 11 million pieces of information our brain processes each second? Most of these remain below our conscious threshold, where we are largely oblivious to their presence.

This second level of activity is taking place all the time, as our subconscious brain scans the environment around us and, more precisely, the incoming signals it receives from it. This is to determine what is important and what is not, and ensure that we consciously attend to anything that needs our attention. Often, things happen in that area below our conscious threshold, which suddenly our brain, or the filter within our brain, decides is something that we need to know about. And so it gets brought up to the level of our conscious attention, and we become aware of it in the more traditional sense of the word. Let me give you an example. Imagine you are busy working away in the office, diligently concentrating on what you are doing. All of a sudden, there is a tremendous crash and a loud bang coming from outside the building. This would capture your attention. You might have been consciously trying to apply attention to something else at that point, but your brain has determined that the activity outside could well be a threat. It is definitely something that you need to be aware of. So although we can choose to voluntarily attend to things, to give them our attention, we need to realize that at the same time there are things outside of our control that can capture our attention at any given moment.

I want to talk to you a bit more about this filter, because I'm sure you've already thought that the key to a lot of what we do in marketing is going to be to get through that filter. If that filter is determining what, of millions of bits of information being processed, actually reaches our conscious threshold, we want to know how that filter works.

It is actually part of our brain called the Reticular Activating System (RAS). It sits right at the top of our brain stem, and it receives information that is coming in from most of our senses (more on that in later chapters). The RAS is receiving all these inputs, all the time, and it is determining what is important to get passed on, and what remains below the level that we are consciously aware of. If you are not currently shouting out 'yes, but how does it decide this?' then I think you need to consider whether you really are a professional marketer! The good news is is that there are some known core factors that drive and shape what the RAS is going to let through. Many of these, as we will see, are linked to our evolutionary past, and would have provided clear advantages to our ongoing survival.

The bad news for us marketers is that some of these factors will be very personal to us. Your name for instance is always going to make it through. Your RAS learned quickly at a young age that your name has the potential to be very important.

Picture the scene... You are in a busy environment, like a networking event. You are talking to people, having a detailed conversation, and shutting out much of what else is going on around you. And yet, you suddenly become aware that somebody else, in a different group, having a different conversation, just mentioned your name. Usually when this happens, we would emphatically deny that we were even listening to their conversation, but your RAS was. It was taking in that input and processing it for possible important information. If it hadn't been doing that, how else could you have 'heard' your name being mentioned? I am confident they didn't have to say it several times to attract your attention. Oh no, usually, your ears 'prick up' at the first mention of your name, and you then move your conscious attention to their conversation to try and establish more details about what was being said about you.

So your name is one element. And we have already mentioned another one – remember the sound of the crash and bang outside your office? Things that have the potential to be fundamental to our survival are another category of what the RAS will let through. But, the brain needs help to identify what has that potential, so essentially it is constantly being vigilant about things around us. The best way we can understand this is to drop down a level, and explore some of the processes that happen below our conscious threshold, to help the RAS determine what could be potentially significant.

Here, we have many possible aspects that can make the cut. Each of these provides the opportunity for us to harness their potential and use them to support our marketing efforts. We can start to include them within our marketing strategies and incorporate them within the content we create. Because, if we know what is likely to enable the RAS to select to attend to out of a very complex, very busy environment, then we've got a really good head start in terms of being noticed.

External cues

The first category of cues we will consider that the RAS is primed to look for and notice are external to us. These are aspects that will be in our environment and happening around us, which convey information that could be vital to our RAS.

Contrast

The first element to consider should probably be the idea of contrast. If something looks different from other things around it, we will consider it subconsciously, and try to determine if it is anything we need to be concerned about. A pen mark on an otherwise plain wall, a house with solar panels on the roof when the rest of the terrace doesn't have them, a yellow 'short-dated' sticker on one punnet of strawberries on the shelf. If it looks different, we need to understand it and so some level of assessment has to go into that comprehension.

FIGURE 3.1 The use of contrast to make specific text stand out

Vivamus bibendum volutpat ornare. d mi feugiat, tristique leo sit amet, luctus nulla. nisi lorem, ac rhoncus mauris laoreet at.

Ut ac velit ipsum. Suspendisse sagittis rhoncus ornare. Nunc quis eros maximus, facilisis nulla, suscipit nulla. Ut condimentum placerat

This is more likely to grab your attention

Mauris luctus sapien augue, ac faucibus tortor iaculis eget. Aenean ut enim mauris.

Size

This is an easy one to understand. Things that are large generally capture our attention easier and faster than something that is very small. There are two main evolutionary reasons why this may be the case. First, in terms of threats, something that is big is likely to be more of a danger to us than something small. However, we also need to appreciate that most of our evolutionary past has not involved considerable volumes of 2D interactions as we do now. Therefore, things that are larger are still interpreted as being closer to us and so command assessment quickly, as their impact could be immediate (pun intended).

FIGURE 3.2 The use of size to make specific text stand out

Vivamus bibendum volutpat ornare. d mi feugiat, tristique leo sit amet, luctus nulla. nisi lorem, ac rhoncus mauris laoreet at.

Ut ac velit ipsum. Suspendisse sagittis rhoncus ornare. Nunc quis eros maximus, facilisis nulla, suscipit nulla. Ut condimentum placerat

This is more likely to grab your attention

Mauris luctus sapien augue, ac faucibus tortor iaculis eget. Aenean ut enim mauris.

Intensity

Some elements may stand out as they are more intense than the rest of the aspects that surround it. If a car alarm is going off, but it is a long way away from us, we may not be consciously aware of it at all. But if it is close to us, and the noise is that much more intense, it actually becomes hard for us to ignore it. In terms of written text, putting something in bold makes it stand out against the rest of the copy, making it more likely that people will notice it.

FIGURE 3.3 The use of intensity to make specific text stand out

Vivamus bibendum volutpat ornare. d mi feugiat, tristique leo sit amet, luctus nulla. nisi lorem, ac rhoncus mauris laoreet at.

Ut ac velit ipsum. Suspendisse sagittis rhoncus ornare. Nunc quis eros maximus, facilisis nulla, suscipit nulla. Ut condimentum placerat

This is more likely to grab your attention

Mauris luctus sapien augue, ac faucibus tortor iaculis eget. Aenean ut enim mauris.

Movement

Again, this is an easy one to correlate to our evolutionary past. If something moves within our external environment we are going to have our attention drawn to it. Is it moving towards us or away from us? How fast is it going? Is it moving smoothly or erratically? We quickly need to identify it and try to make sense of it, so it is going to demand additional processing.

FIGURE 3.4 Incorporating elements suggesting movement into a design

MATTERS

Novelty

What is new? What is different? What has changed? It now appears as though the brain uses predictions a lot more than we had previously thought in order to efficiently navigate our way through life.[1] So, things that don't fit the prediction, that weren't there before, or that appear out of place, will again be in line for deeper processing. Do you remember those children's puzzles where you had to spot 10 differences between two, virtually identical images? This plays to the natural talent our brains have for noticing changes that exist around us. This could be changes in our physical environment or changes in the people around us. Which leads me on to the next point.

FIGURE 3.5 The use of novelty to make specific text stand out

Vivamus bibendum volutpat ornare. d mi feugiat, tristique leo sit amet, luctus nulla. nisi lorem, ac rhoncus placerat Mauris luctus sapien augue, ac faucibus tortor iaculis eget. Aenean ut

This is more likely to grab your attention

maximus, facilisis nulla, suscipit nulla. Ut ac velit condiment um

Emotions

If there are faces around us, we will seek to understand their emotions as a result of interpreting facial expressions. If they are intense, or contrast with other people around us, we will again be compelled to try and understand why.

FIGURE 3.6 Incorporating faces and potentially emotions into a design

Colour

As we will explore in more detail later, colour has an effect on us, which means it too will naturally affect our levels of attention. Some of those are based on the environment around it and the creation of contrast, for example, but we can also see that some aspects of colour seem to naturally appeal to us, they seem to naturally pull our attention in. Take red, for example. In our evolutionary past, the colour red in our natural environment would be something that we needed to attend to. It may be bad news, such as blood or potentially dangerous insects, or it could be good news in terms of ripe fruit and berries. Either way, we tend to notice red.

Request it

Finally, the last way for us to get attention is... to ask for it. It is a very simple technique and is particularly effective if you're doing pitches, presentations, public speaking or anything involving an audio element. Just ask your audience for their attention. It is amazing how effective that can actually be. Maybe we are all still affected by the requests made to us from the front of the class after all!

Internal cues

So, there we have a range of different devices we can easily incorporate in order to get ourselves processed at some level. But we also need to acknowledge that some of the aspects that trigger the RAS will come to it from internal information. So, let's have a think about what some of those could be.

Interest

One of the most differentiating aspects will be things that are of interest to us. If you have a particular interest in ice hockey or old farm machinery or gluten free recipes or upcycling furniture, you are going to notice elements that reinforce it or that are related to it in some way. Similarly, if you are currently grappling at work with an issue about how to increase ROI on digital campaigns or utilizing AI to optimize automation or launching an existing product into a new sector, you will attend to content referring to these. If there are topics that are interesting to us at an individual level, they will naturally capture our attention.

Emotions

I know what you are going to say: 'But Katie, we've already looked at emotions'. Well, yes, we have. Well done for noticing that. However, this is not about the external presentation of emotions, this is about the audiences' internal emotional state. What speaks to them at an emotional level? How are they likely to be feeling at the moment? And what resonates with that emotional experience that they are currently having? Anything that aligns with that, or that reaches out and connects to us on an emotional level, is going to capture our attention. It's going to be significant and potentially really important for us.

Organic

We also need to accept that we have our internal, organic states, which will also affect our processing of the environment around us.

If you are hungry or thirsty during the last conference session before lunch, the attention you give to that presenter, that content, that part of the agenda, is going to be waning. It's going to be slipping away. The same with other vital internal states such as pain or tiredness. These are all going to be factors that majorly influence what catches our attention, as well as affecting how much attention we've got to give.

Train of thought

This is where I think the RAS most clearly demonstrates its existence. Imagine you have been thinking about something like buying a new car. You may have been doing your research and narrowing down your options. You may even have been paying more attention to noticing other cars on the roads as you travel around. But then, as you make your final decision for, say, an Audi Q3, you'll suddenly start to notice lots and lots of them on the roads around you. Now, they were always there, it is just that you didn't pay them attention because your RAS didn't think they were important to you. Whereas, as soon as you start having a train of thought about it, your RAS learns that they are significant, and so it highlights for you all the relevant evidence, information and examples.

Effort

The final internal aspect for us to be aware of is the consideration of how much effort is required. This is something that we will come back to, again and again, because one of the key things our brain does for us is make a very quick assessment of how much cognitive capacity is required in order to complete a task. If it decides it is too much, it won't apply the conscious attention that is required. So, we are much less likely to engage with very busy, complicated text for instance. Our brains will probably just deem it too much like hard work. It won't make it through the RAS and so consciously we're not going to be able to get that same level of involvement that we might seek.

Holding attention

This moves us neatly into the final area I want you to start considering under the topic of attention. Are you still with me on this? I hope so!

As marketers, we have so far been looking at the different devices and methods we can adopt to attract the attention of our audiences. If we want the RAS to notice us, to get the information into that precious conscious area where cognitive resource is applied, we've got to capture that attention in the first place. So it is only natural that we should spend the majority of our time exploring this. However, we also need to ensure that we then *hold* their attention once we have it. And, as I am sure you are aware, this is no mean feat. Our audiences are becoming faster to judge, quicker to move on and as a result less tolerant. There are lots of challenges to holding attention, because there are lots of demands on our attention all the time. But if we want to get a message across, if we want someone to understand our core values or to complete a call to action at the end of a piece of content, we've got to retain their attention. And as we shall see in later chapters, retaining their attention is a completely different challenge, which we approach using some complementary, but different tools. For now though, if you want a masterclass in how marketing has achieved this concept in the past, look no further than the Johnnie Walker advert they ran in 2009.[2] It is over six mins long – yes, that is not a typo. Over six mins long. And it was filmed in one continuous sequence. If you are not already familiar with it, give up six minutes of your life just to see how it can be done in some very simple, understated ways. I think you'll agree with me, the result is actually quite compelling.

We have considered aspects that *capture* attention and briefly been reminded of the need to *hold* our audiences' attention. The final concept of this trilogy is ensuring that we actively *direct* our audiences' attention too. We will explore how we can potentially manage the way people engage with the materials and the content that we are presenting. You may have seen examples like this as they crop up on social media from time to time. I'll just give you a moment to follow the natural direction that you are guided through in Figure 3.7.

FIGURE 3.7 Creating designs to direct the reader's attention in a specific order

You will read this last
You will read this first
And then you will read this
Then this one

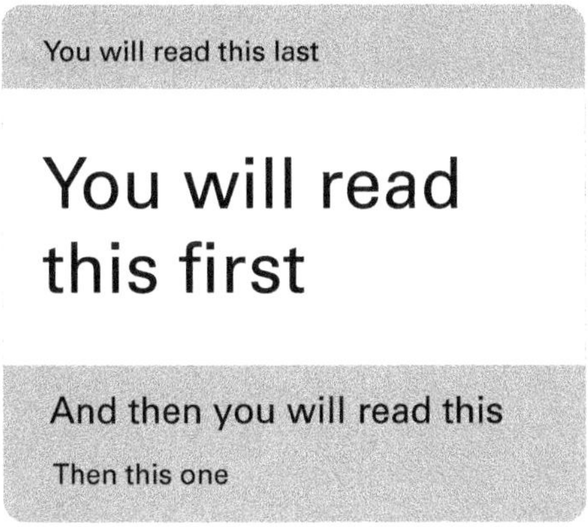

I love this, because if I'd got some eye-tracking goggles on you, it is so easy to predict where your eyes would fall during this. But I am jumping ahead, this is content for future chapters. For now, I want to congratulate you for paying attention all of that time, and also thank you for it too. I know it will not have been easy, but I assure you, it will be worth it.

Chapter summary

1 There is not just one sort of attention – it can in fact be sustained, alternating, divided or selective.

2 All models of attention acknowledge some degree of filtering – what you do pay attention to, and consequently what you do not.

3 Your Reticular Activating System (RAS) is the filter that determines what you actually pay conscious attention to.

4 There are predictable external and internal aspects that the RAS will be more likely to apply conscious attention to.

5 Many of these aspects have an evolutionary or survival basis to them, which is hardwired into us.

6 Your attention can be applied consciously or taken involuntarily by what your RAS decides is more urgent.

7. Separate consideration should be given to ensuring we can hold our audiences' attention, as this is becoming increasingly challenging.
8. Your marketing approaches need to individually and intentionally approach capturing, retaining and directing the attention of your audience.

Notes

1. Clark, A (2023) *The Experience Machine: How our minds predict and shape reality*, Pantheon Books, New York
2. Johnnie Walker (2009) The Man Who Walked Around the World, YouTube, www.youtube.com/watch?v=fZ6aiVg2qVk (archived at https://perma.cc/Y3DW-BVTJ)

4

It's going to get emotional

Context

For most of our evolutionary past, we humans have believed ourselves to be highly rational creatures, who also experience emotions. We have believed this rational facility is what raises us above the level of other animals we share our planet with, and we have educated, celebrated and commemorated our prowess in this field. However, discoveries in neuroscience have shown that we are in fact not driven by rational thoughts and decisions, but by emotional ones. So if we want to know how to understand, reach and connect with our audiences, we need to do so at an emotional level.

If we take a moment to look back through much of our history, we can see that we have invested inordinate amounts of time, effort and energy in wanting knowledge – seeking scientific clarity and wanting to understand things and be able to explain them in very logical, meticulous, rational steps. Whether those 'things' were ourselves and our fellow humans or the environments we live in. Whether we were using reductionist approaches to comprehend the cells and atoms that make up our world or using laws and models to understand vast concepts such as astronomy and our position within the stars. Scientific discoveries and advances were sought, debated, published and disseminated. And to a great extent, they still are. The irony is not lost on me that the insights I am sharing with you in this book have been researched, discovered, verified and shared by adopting the same robust scientific principles.

I understand much of where this has come from. If we believe ourselves to be rational, then we become much more predictable. We can start to understand the way people will behave in certain situations, and what we might need to do in order to change those behaviours. But, you see, we do not always do this. Economists have struggled with how irrational we can be at times, and how easily and often we will act against what rationally makes the most sense.

Consider the following example, which has been much studied in neuroeconomics – the Ultimatum Game.[1] The game involves two players – do you mind if for the purposes of illustration I make that me and you? Thank you. So, the game will ask us to imagine that I have just been given £100, and I am going to share it with you. I have one chance to make you an offer of how we will split the £100 I have just received. You, as the respondent, can either accept or reject the offer I make. If you reject the offer I make, we both go away with nothing. If you accept it, we both go away with the share of the £100 that we have agreed to. Is that clear? We have one chance at this, and you get to decide the outcome.

So, as the proposer, I may consider my options, try to predict your likely response and ultimately come up with an offer for you. In this instance, my offer is that I retain £80 and I will give you £20 of what I have just received.

How will you respond to that offer? Do you choose to accept it or reject it?

Now, rationally, we might think that it makes sense to accept that offer. After all, you came into the room with nothing, and if you accept it, you get to leave £20 better off. However, when this experiment is conducted for real (rather than just in your imagination as it currently is), we find that most responders will actually reject the offer of £20 that has been made to them. Why? Good question. Why would you do that when it is clearly the action that leaves you worse off? Well, you would do it because this action allows you to do something else too. Rejecting my offer allows you to punish me (the proposer) for making such a derisory proposal to you in the first place. And in the majority of instances, this emotional response is more compelling to us than the rational outcome we may previously have predicted. We actually prioritize the emotional need to retaliate, find a more 'fair' outcome and maybe even retain some dignity over our individual economic benefit.

Building on this are years of studies that show us that we humans do not make rational decisions. The decisions we make are motivated by our emotions, and so, in turn, our behaviours are driven by them too. This gives us marketers a significant opportunity. If we can reach our audiences on an emotional level, we will be connecting to the decision-making parts of their brains. So, where and how can we do that?

An experiment on discounts

Consider the examples in Figure 4.1 which I presented to a number of participants whilst they were wearing an EEG headset.

Rationally, they are all exactly the same offer. But the way we respond to each of them is very different. Which do you feel is the most compelling or the most motivating for you?

If you are like the majority of participants who completed this research for me, you may have said that the '50% off' one appealed to you the most. When we look at the brain activity that is taking place when people view this offer, we can see that of the four options, this is the one they are most interested in, most focused on and even most excited by. However, it also scored the second highest for stress – that is our assessment of the cognitive capacity required – suggesting that it created a feeling of overwhelm in the observers. Interestingly, this was significantly worse for the male participants, who presented an average score that was 18.26 per cent higher than the average score presented by the female participants on this metric.

FIGURE 4.1 Examples of different sales promotions, offering the same rational discount

So, we can see that even items that we may feel are innocuous and removed from emotional responses, such as sales tags, actually trigger us to react in very different ways.

Emotions trigger actions

The other exciting aspect of this is that emotions trigger us to act. Full stop. If we want our audiences, either internal or external, to respond or engage, to be motivated to change their views or even their behaviours, we need to connect with their emotions. Simply presenting data to them is not going to work. And yet, this is what we find is the case around the world in boardrooms and on conference stages, in the media and online, in promotional literature and sales conversations. Such a wasted opportunity.

You see, I have known for many years now that I need to eat five portions of fruit and vegetables each day. I know that. It is a fact. It is data. But do I do it? No, not consistently. You see the data has not connected with me, it feels removed from my situation and priorities. However, as soon as I am told the story of someone who is similar in age to me, who has a busy husband and children, who runs their own business, and who previously struggled with tiredness and fatigue, then I am listening. The more emotional the language, the more I am drawn into their situation. Now, if you tell me that their life has changed as a result of consistently eating these same five portions of fruit and vegetables each day, and you use emotions to describe those changes to me, I am on the journey with you. I can almost taste the vitamin C and the fibre from here!

Emotions engage us and they persuade us. Data does not. There is a role for data though, I am not in any way dismissing its relevance. However, it needs to become part of the message, a part of the evidence that backs up examples and stories that have emotional content. Using data can tell a message, but it is less likely to change people than emotional content is. The inclusion of emotionally charged storytelling, case-studies, testimonials and content will do far more towards changing behaviour than data ever will. So, never be afraid to bring emotions into the boardroom.

In fact, the boardroom should thank you if you do. You see, incorporating emotions into marketing strategies has proven to be highly profitable and effective at driving unparalleled business growth.[2] By addressing some of the 'emotional motivators' that audiences have, brands and organizations can more than double the lifetime value of a customer, because they will:

- consume more of your products/services
- buy more often
- be less sensitive to price
- give your marketing and communications more of their precious attention
- trust your recommendations and suggestions
- refer you more often[3]

I cannot imagine any boardroom in the world where these results would not be welcomed with open arms.

Now I don't know how much you currently think about emotions as you're going about your marketing functions. Whether that's pricing, messages, campaigns, projects, research – to what extent do emotions currently get considered as part of the decisions that you make? Well, whatever your answer to that is, I am hoping that as a result of reading this section we are going to change things for you. Because I want you to really see that emotions hold tremendous power for us, they can give us enhanced access to our audiences and target audiences in ways that potentially we may not have considered or been able to achieve before.

So, let's understand why that is.

What are emotions?

The logical place for us to start is by exploring what emotions actually are and how they are triggered and processed within the human brain.

When we seek to define what an emotion is some key characteristics frequently crop up. These usually refer to something along the lines of 'subjectively experienced conscious mental states' which 'vary in intensity', and are 'brought on by our thoughts, or people or events around us'. So, this is a good start – these tell us some interesting aspects, don't they?

First, they tell us that not everyone experiences the same emotions. This may be because we experience them at a different intensity or because we actually experience completely different emotions. Imagine you and a colleague are just about to present a report to your senior team. On the way to the room, one of you may be feeling excitement and anticipation, whilst the other could be feeling fear and even dread. How could these be so different? Well, essentially it is because you have different thoughts about the upcoming situation. The one feeling excitement and anticipation may be feeling that way because they are optimistic, confident and are keen to show your expertise and insights to your senior managers. On the other hand, the one feeling fear and dread may be feeling that way because they have had bad experiences in the past or because they know they have not prepared properly for this. The same external trigger can cause very different emotions.

Similarly, the intensity of emotions can vary between individuals too. Something that one of us finds hilariously funny could be mildly amusing to another person. And a scene in a film one of us finds unbelievably poignant and tragic may for someone else be just a bit sad but mainly unnecessarily long (yes, the bridge scene in *Meet Joe Black*, I'm talking about you).

Interestingly, our emotional responses appear to have a universal element to them, suggesting that they are somehow built into us. The late Dr Paul Ekman is in many ways regarded as the major force behind our understanding and our appreciation of emotions. He dedicated his career to understanding emotions, and creating systems by which we can measure and analyse them. As part of his work in this area, he travelled the world studying emotions in different groups of people, in extremely diverse and sometimes remote populations. As a result of this, Dr Ekman established some fabulous and now famous insights, which have themselves gone on to trigger more research and applications in many different arenas. One of his most influential findings was identifying that there are seven universal emotions.[4] Now, by 'universal' I mean anywhere in the world that he went or that we may go, these emotions will be presented in the same way. It is important to acknowledge that what triggers them may be very different though, and vary greatly depending on where we are. So, if I am experiencing happiness here in the Western world, what triggers me to feel happy might be very different from somebody who lives in a tribal community in South America for instance. Their trigger for happiness may be very different from my own. However, when we are triggered, our expression of that happiness will be universal – it will be exactly the same. We will both be smiling or even laughing and present the same configuration of muscle movements on our faces.

Regardless of the events and actions that trigger emotions, there is always a degree of thought involved too. It is what I *think* about it that will affect how I feel about it. Do I perceive the dog running towards me as being cute or scary? Do I empathize with the plight of the homeless person sitting in the doorway or not? Do I feel comfortable with the colour of my teeth at the moment or not? Each of these will prompt a different response in our brain, and in turn that will start to create physiological changes within it, and indeed, throughout much of our body too.

Emotions within the brain

The part of our brain that plays the most dominant role in emotions is the amygdala. This small almond-shaped area (which is where the name 'amygdala' comes from – the Greek word for almond) is involved in both positive and negative emotional responses. However, it is particularly adept at some of the more negative ones, as it has a vital role to play in triggering the 'fight or flight' response which is part of our key survival functions. Therefore, the amygdala is perhaps most important when it comes to activating emotions such as fear, anxiety and stress. This is typically done in response to situations that feel dangerous or threaten our physical or psychological wellbeing in some way.

The amygdala takes in information from our senses, and from some of our physiological and cognitive processes too (such as digestion and memories), and processes it all very quickly to produce an appropriate response. This response will then trigger a number of changes within the brain, such as the release of specific neurotransmitters and hormones as we have previously seen. These, in turn, affect changes throughout our bodies, such as diverting resources or changing physiological processes. And it does all of this exceptionally quickly, and without us really knowing about it until afterwards.

Just think about the last time someone (or something) made you jump. You know, when something really unexpected happens like you walk into another room, and someone is innocently standing just behind the door? Your amygdala will get to work within less than 80 milliseconds, assess the threat level and trigger changes before we have consciously worked out what is going on.[5] That shock response has already set your pulse racing, your rate of breathing will increase and your blood flow changes to support you as you prepare to either fight or flee. But do you remember that it also affects the way your brain works? The cerebral cortex, which as we know is involved in our reasoning, judgement and decision-making processes, becomes impaired. We are therefore less likely to be able to think clearly or make good decisions when we are in this state, and so our responses may often be out of proportion to the scale of the threat being experienced. Emotional responses have happened before we have had any chance to process or compute what is going on. Emotions happen first, and rationality then has to try and catch up.

One final aspect of emotions that you need to be aware of before we can begin to explore how to best utilize these insights in our marketing approaches is the extent to which we are hardwired to respond to them.

For most of our evolutionary past, it has been useful to us as human beings to be able to understand how other people are feeling.

This is why our experience of emotions is literally written all over our faces. It helps us to be able to determine whether someone who is approaching us is angry or feeling more convivial. It changes the way we treat people if we can see whether they are upset or scared. And it indicates our next course of action – for example if we can see someone is presenting disgust or pleasure on their face after just having eaten something. Being able to support, nurture, defuse and even avoid people, depending on their emotions, is a valuable survival adaptation, which has been passed through hundreds of generations. And it still serves us today.

Only now we may abbreviate it to a simple 😃.

But again, much of the processing of these cues regarding emotions will happen below our conscious threshold. Remember those 11 million pieces of information our brain processes every second? Well, many of those will be used to interpret and understand the emotions of those around us. Reading the nuances of their facial expressions, vocal patterns and body language will provide information and vital insights which our amygdala will use to identify and produce the response it believes is the most appropriate. This is why we should learn to trust our instincts more!

You see, unconsciously our amygdala may pick up on subtle cues that give away when someone is lying or uncomfortable with what they are saying. We may realize that someone's anger is actually based in fear rather than hatred. Or we may see the pain we have caused someone, despite how hard they are attempting to conceal it. And the realization of each of these will again trigger changes in us, which we may not be able to articulate or identify. But the changes will happen anyway.

> The processes we have for quickly reading and attributing emotions are not limited to people either. I remember when I did some work on behalf of Honda back in 2008: they had spent a number of years developing a humanoid robot referred to as ASIMO. It was about the height of an average nine-year-old child and had been developed to recognize and interact with a wide range of objects and even gestures within its environment. However, its construction and responses were so accurate

that it was hard for all of us to always refer to ASIMO as an 'it' and not as a 'he', particularly as this was before the days of people openly declaring their pronoun preferences. My brain must have processed the way this robot functioned and combined it with some nurturing tendency relevant to a nine-year-old child, and the combination meant that I (like many others) started to attribute emotions to ASIMO, and if we weren't careful, fully fledged anthropomorphism would soon take hold. If you have not heard of ASIMO before, go and look it up, maybe then you will understand the challenge we had!

You see, some of these drives are so powerful within us that we can start to look for and see them even where they do not exist. Welcome to the world of pareidolia. You may not have heard the word before, but I wouldn't mind betting that you have experienced it. This refers to the way our brain 'sees' meaningful patterns and shapes in ambiguous or random visual objects. For example, have you ever felt as though you can see a face in the surface of the moon? Or maybe in the formation of a car's headlights and radiator grill as it approaches you? Or the face of Jesus on the inside of your Marmite lid (yes, Google that one, it really happened). So powerful is our need to identify faces within our visual environment that we have an area of our brain called the fusiform gyrus, which is dedicated to doing just this. Moreover, it will help us to determine whether the faces are familiar and known to us or not, and finally it helps us to read their expression. It is permanently looking for the configuration of shapes that makes up a face, which is why we sometimes activate these processes even when there is no human basis behind them.

If this superpower is working away within us at all times, and also within our audiences too, does it not make sense for us to utilize this information to help us get noticed, promote our core values and convey our relevant messages?

Being intentional about harnessing these natural processes will forever change the way you approach your marketing. If we can understand the way someone is likely to respond to our content or messaging, and the emotional activation that goes with that, we can connect with our audiences in new and exciting ways.

So, where do we start?

First, we need to acknowledge that there are two considerations here. There are the emotions we present going out from our organization and there are the emotions we understand will exist in our audiences. Hopefully, if you have been paying attention thus far, you will realize that the two are not entirely unconnected. The emotions that we present going out from our organizations will have an effect on the audiences who receive them. They cannot help it. But, similarly, the emotions that they may be experiencing at any given time will also affect the way they perceive and respond to us and our materials. Hmmm, not so easy then!

Well, let's park that thought for a few paragraphs and start with the areas that are easier for us to influence – those emotions that we present as an organization or brand. There are a number of ways that we can present emotions. They can essentially be summarized as follows:

1. the brand itself
2. the language we use
3. the images we use

1. The brand itself

Does your brand have a personality? Does it have character and characteristics that make it literally come to life? If not, why not?

You see, there are no areas in our brain that are designed for the processing of brands... yet! In terms of our evolutionary past, brands are relatively recent. So, the way we notice them, react to them, organize them, recall them, connect with them and make decisions about them all follow processes that our brain has but which were designed for other purposes.

For example, when we first see information pertaining to a brand, such as its logo, we form an impression of it. Depending on the colours, words and devices used, and the location where it appears, the brain will become activated in a variety of unconscious ways as it processes this information and reaches conclusions about what it has seen. However, the interesting aspect of this is that the next time we are exposed to the same logo, or details relating to the same brand, we attach that information to the previous information, and something called a schema is created. Simply put, schemas are the brain's way of organizing information that is all related. So, if I ask you to think about a dog, you will automatically come up with one based on the schema within your brain. This might be based on one you grew up with or maybe one you have now. Your attitude towards whether dogs are pleasant and fun or aggressive and dangerous will depend on the information you have within your schema. This information is gathered unconsciously over the course of your lifetime. Small pieces of information are constantly being filed away, helping us to make sense of our environment and experiences.

And so it is for brands. We create schemas for those too. We assimilate all that we know and believe to be true about brands into dedicated schemas, which then shape and influence the perceptions we have towards them and the decisions we make about them.[6] However, at their usual level, this unconscious information is unlikely to create powerful connections and contain factors that motivate us to act.

So, in order to allow our audiences to connect with our brands and have feelings towards them, the easiest step for us to take is to present those brands through personification. Consider the approach taken by Aldi.[7]

EXAMPLE

As a brand, Aldi is known for providing lower cost imitations of many well-known brands. This is particularly apparent in how they package and present their products, which are designed to very closely mimic the key identities of the original brands themselves. But along with this, Aldi have developed a real personality behind their brand, and this is particularly evident in their social media content.

Through social media Aldi has become known for being cheeky, humorous and more than a little disruptive.[8] Yes, they have very clear messages that they consistently convey regarding their levels of quality and value, and they are proud to call out others who do not meet their priorities. However, they also create a fabulous personality by applying two more aspects. First, they are reactive. They respond to their customer messages and share their responses (which are sometimes not very complimentary to their customer). This creates a sense of conversation and accessibility that starts to give the brand its personality, but also conveys emotional elements throughout. An additional aspect of this reactivity is that they are quick to publish content on current trends and topics. This suggests a very dynamic process behind their content creation which shows them to not be restricted by layers of approvals and sign-offs (which, as we all know, can take a prohibitive amount of time). This leads me on to the second aspect they so brilliantly adopt.

Knowing their reputation for being like David against the Goliaths of the grocery industry in the UK, they have shown themselves to be innovative, dynamic and quite self-deprecating too. Using images from current news items but applying them to their own position, performance and history is a fabulous way to show their personality, and create a loyal following who actually have emotional connections with the brand. A low-cost supermarket! And all of this without even mentioning their iconic use of Kevin the Carrot for their Christmas campaigns since 2016.

Other brands who in my opinion have created very strong emotionally driven personalities for themselves include Red Bull (boundary pushing dare-devils), Patagonia (passionate environmentalists) and Nike (building inspiration and self-belief). But do you notice how these examples all come from the world of B2C? Sadly, as yet, no-one is really daring to create such strong emotional connection with brands in the B2B space. And yet our brain does not differentiate. Our brain does not change the way it processes information or pays attention, depending on whether it is at work or not. So, let's see more of these aspects being presented in B2B organizations, I assure you, there is a lot to gain for those pioneers who do embrace it.

This brings me on to the second element of how we can include and present emotional content. Language.

2. The language we use

Consider the difference between the following two passages of text.

> **A.** More than 120 people have been killed after a powerful earthquake struck a remote region of Tibet on Tuesday morning, with tremors felt across the Himalayas in neighbouring Nepal, Bhutan and parts of northern India. According to the United States Geological Survey (USGS), the 7.1-magnitude quake struck at 9:05 am local time at a depth of 10 kilometers (6.2 miles) and was followed by multiple aftershocks. The energy unleashed by the tectonic movement toppled houses in remote Himalayan villages, rocked a nearby Tibetan holy city and rattled visitors to a Mount Everest base camp. *(Adapted from* https://edition.cnn.com/2025/01/06/china/china-tibet-earthquake-intl-hnk/index.html*)*

> **B.** People's lives were ripped apart in Tibet yesterday as a powerful earthquake interrupted their morning routines. In the town shops, items began cascading off the shelves and people abandoned their baskets in the aisles and ran, screaming, out into the streets. In remote villages, homes began to crumble, as families desperately called out to each other, and ran into heartfelt embraces. They could only watch on in horror as their homes were reduced to rubble, all their possessions buried and their worlds turned to dust. With no warning, security was lost, homes were lost, and lives were lost. But before they can grieve, they must try to save what they can of their possessions and start to find materials to build again.

Which one moves you? Which one do you literally feel your body responding to? Which one is more likely to make you want to donate?

Conveying emotions within our language will have a significant effect on the audience who is reading, watching or listening to our content. Factual statements will not. They may provide greater clarity or be more objective but they are less likely to bring about any change in our attitude, approach or behaviour. Only emotional language can do that.

So where and how can you incorporate this into your communications? The only answer is... everywhere. From the emails you send out to the scripts your chatbot uses. From your social media posts to your printed materials. From your sales teams at a networking event to your tagline on vans. Emotional language will engage and connect people to your brand, and it will do it quickly too.

3. The images we use

Even faster than language, though, is our use of images. It is widely accepted that the brain processes images faster than text, although precise details on how much faster vary from study to study. If you want to capture the attention of your audience, or convey some key emotions to them, images are the fastest way to do that.

Using human faces within marketing campaigns can provide a very efficient short-cut to influencing our audiences. The right expression, on the right face, will not only get you noticed but it could create a variety of responses – empathy, disgust, curiosity, joy, fear, sadness, desire, regret, inspiration; the list goes on.

However, it comes with a warning. Please, just remember how good we naturally are at reading these emotions. This is something we have been hardwired to do since birth, and it is processed below our conscious threshold most of the time, so we may not be able to establish or articulate the interpretations we have made. The reason this needs to have a warning attached is because far too often I see organizations and brands using images that are not conveying what they think they are.

Do you remember how our true feelings can 'leak' out in our expressions? Well, this is also true of people who are featured in the images and videos we use. Consider the very common situations of someone having their headshot taken to be used on the website or go out with a press release, or someone else being asked to record some content on video for social media. If either of these individuals are uncomfortable about what they are doing, it is very easy for that to show in their facial expressions and body language. Even the images we can purchase from reputable online resources feature models who are clearly not authentic in the emotions they are trying to convey. And our audience will subconsciously pick up on this, even if to our rational eyes the smile or the expression looks genuine. They will be aware that someone is unhappy or incongruent or disingenuous, and that may trigger a 'gut feeling' that something is off. Such a feeling can be very damaging to your brand and, ultimately, profits.

Instead, be sure to use authentic, genuine images, where you have established the emotional intention behind the shots being taken. Use confident, sincere people in your videos, and try to let them speak their own words, rather than using an artificial or forced script.

Building on this aspect, we need to return to the idea of how our audience may already be feeling, and what affect (or opportunities) that has on (or for) us.

Scales of emotion

When we explore the way someone experiences an emotion, there are three key scales we tend to refer to – pleasure, arousal and dominance.[9]

- **Pleasure** refers to whether the emotion is enjoyable or not, so happiness scores highly on this scale, but anxiety has a low score.
- **Arousal** refers to whether the emotion energizes us or not, so anger may score highly on this one, but boredom has a low score.
- **Dominance** refers to whether the emotion makes us feel in control or not, so assertiveness would score low here but fear has a high score.

Understanding where an emotion is on each of these three scales gives us vital information about how we need to approach and support someone who is experiencing that emotion.

If we take the example of someone who is feeling disgusted, they will have very low levels of pleasure (because they have just experienced something unpleasant), very high levels of arousal (because we are very motivated to remove the offending stimulus from being in our vicinity) and average or medium levels of dominance (as we do have some power, but also something bad has just happened to us).

The really interesting element in this approach is when we then turn this around and consider what we need to do in order to communicate effectively to this person. If they are unhappy and aroused, they are not going to be paying us much attention at all, so we really need to work to bring their arousal level down, and let them return to a calmer state, before anything we show or present to them is going to connect.

This is important for us to understand, because we now have sufficient technology to enable us to start to use emotions as part of our segmentations. I am sure you have come across DOOH (Digital, Out of Home) boards. These are the digital boards you may see on the high street, in your local cinema, gym or coffee shop, etc, for which you can purchase air-time to promote your marketing messages. They have a variety of changing digital content, which they present for a short period of time. Now, some of these boards are equipped with cameras on them which enable the boards to 'look' around them and capture information about the people who are in the immediate vicinity. This information is not sensitive, and the cameras do not record what they see, so there are few legal issues here. However, the boards use this basic data to work out the demographics of the audience in front of them. This is then used to select the most appropriate content for that audience at that time. So if, for example, a brand of deodorant were to use one of these boards, they could set them up to show content targeting either male or female consumers, depending on the dominant gender of the people in the vicinity of the board. Clever, eh?

But wait, it gets better. Some of these lenses now incorporate AI to enable them to 'read' the facial expressions of the people near to them too, and from that determine their likely emotional state. So, they can literally 'read the room' and then select and present the best content that will be most likely to resonate with the demographic and the dominant mood of the people in front of it. How fabulous is that? So the advertising content can be selected based upon the prevailing mood of the people who will receive it. No more adverts presenting images of smiling, happy, care-free people, when the majority of the audience is concerned about the cost-of-living crisis or the terrible event in the news this morning. What fabulous opportunities these emotional connections now allow us to make.

Chapter summary

1 We are not as rational as we think we are – emotions drive our thoughts, feelings, decisions and behaviours.
2 Different people experience the same emotion differently, and are potentially triggered by different things or events.
3 There are seven universal emotions, which are always presented via the same facial expressions, regardless of where you are in the world.
4 Emotions are processed by the amygdala, which is part of the ancient limbic system of our brain.
5 Being able to understand and interpret the emotions of those around us has been of great value and contributed significantly to our ongoing survival.
6 Brands can display emotions through their content, images and overall strategies.
7 Considering the PAD score of emotional experiences allows us to understand more about their origins and opportunities.
8 Emerging developments in technology now enable digital content to be selected and presented, based on the emotions of the audience.

Notes

1 Güth, W, Schmittberger, R and Schwarze, B (1982) An experimental analysis of ultimatum bargaining, *Journal of Economic Behavior & Organization*, 3 (4), 367–88, https://doi.org/10.1016/0167-2681(82)90011-7 (archived at https://perma.cc/Y5PL-S2NF)

2 Magids, S, Zorfas, A and Leemon, D (2015) The new science of customer emotions, *Harvard Business Review* (November), https://hbr.org/2015/11/the-new-science-of-customer-emotions (archived at https://perma.cc/9D4X-X52T)

3 Zorfas, A and Leemon, D (2016) An emotional connection matters more than customer satisfaction, *Harvard Business Review* (29 August), https://hbr.org/2016/08/an-emotional-connection-matters-more-than-customer-satisfaction (archived at https://perma.cc/8NUC-QCRT)

4 Ekman, P (1971) Universal and cultural differences in facial expressions of emotions. In Cole, J (ed), *Nebraska Symposium on Motivation* (1971), 19, 207–83, University of Nebraska Press, Lincoln

5 Méndez-Bértolo, C et al (2016) A fast pathway for fear in human amygdala, *Nature Neuroscience*, 19 (8), 1041–49, https://doi.org/10.1038/nn.4324 (archived at https://perma.cc/BF5X-RGFM)

6 Halkias, G (2015) Mental representation of brands: A schema-based approach to consumers' organization of market knowledge, *Journal of Product & Brand Management*, 24 (5), 438–48

7 Aldi (n.d.) What's so special about Aldi?, www.aldi.co.uk/corporate/about (archived at https://perma.cc/M7C4-J9GS)

8 Marketing Week (n.d.) How Aldi reached millions to achieve social media fame, www.marketingweek.com/aldi-social-media-fame/ (archived at https://perma.cc/JTQ6-DDBG)

9 Mehrabian, A and Russell, J A (1974) *An Approach to Environmental Psychology*, MIT Press, Cambridge

5

Reasons to remember

Context

It's really important for us within marketing that we not only get noticed but that we also get remembered. In most instances, your prospects are unlikely to be ready and in buying mode when they first encounter you. So, you need to enable them to form an impression of you, your organization, your brand, your products and services, that is enduring. It needs to last. You want that impression to become a memory, which will return to them in the very moment when they are considering making a purchase.

In this chapter, we are going to spend some time learning about memory and the different sorts of memories we usually have. Too often, people think and talk of 'memory' as being one entity – either we have a good memory or a poor memory. However, this is not the case at all. There are actually many different sorts of memory that we need to understand and appreciate. They can be broken down according to whether they are:

- **Factual** – things that we *know*
- **Skills** – things that we *show*
- **Experiences** – times that are *retro* (... what? It helps me to remember them all ok?!)
- **Duration** – defined by length of the memory, how *long ago*

Let's explore and expand on these, so that we can be more precise and specific when we are considering the types of memories we want people to be creating, and therefore how we can go about it.

Explicit memory

Sometimes referred to as declarative memories, explicit memories are those memories that we can talk about, we can articulate, we can express... explicitly. They in turn fall into two quite distinct categories: episodic and semantic.

Episodic explicit memories (or episodic declarative memories) are the memories we have that relate to events and our personal experiences. We are able to talk about them, we are able to share our experiences and we are able to recount them to other people. For example, maybe you can recall your last birthday celebration. You may be able to remember where you celebrated it, who you were with, you might even be able to remember what you ate or what you wore. Or maybe you can recall a conversation that you had with somebody yesterday, maybe with a friend or a colleague, and recall the details of what that conversation was actually about. These are examples of events and experiences that you've had, which you can explicitly share, declare and talk about.

Semantic explicit memories (or semantic declarative memories) are not about events or episodes but more about knowledge. These are things that we know. We may know that Rome is the capital city of Italy. We may know that water boils at 100°C and freezes at 0°C. These are facts that we know, scientific facts, statements. We can also know things like the rules of chess, so we know which direction any of the pieces is able to move in. These kinds of memories, semantic explicit memories, are things that we know, but they are removed from any particular event or experience. For example, our knowledge of the ways the pieces of chess can actually move is knowledge we have independent of any memory we might have of perhaps being taught to play chess by a beloved grandparent. Knowing that Rome is the capital of Italy is knowledge that we have independent of any experience we may have had of ever going there. These sorts of memories are facts that we have learned and come to know, and again they are memories that we can share and declare.

Now we need to consider the opposite – the forms of memories that we cannot state, articulate or explicitly share. These are referred to as implicit (or non-declarative) memories.

Implicit memory

Again, these fall into two distinct categories: procedural and priming.

Procedural implicit memories are essentially the skills that we have, the habits that we've developed, the things that we are able to do. Maybe you can ride a bike or swim or play a musical instrument. Even if you have not done these things for many years, if you've learned to do them to a competent level before, the chances are you could make a good effort at it, and you would quickly reconnect with your previous skills in that area again. You may not be as fast, stable or good for a start, but it is not as if you have to go all the way back to the start and learn the whole process all over again. Those skills have stayed with you. That skill is within you, so you are quickly able to effectively dust it off and resurrect it when the need arises. When you first learned the skill, there may have been great complexity and effort required in order to achieve it, and yet, when you become proficient at it, the same outcome can be achieved but with very little conscious effort at all. Does this sound familiar? Does it remind you of the heuristics you learned about earlier on? I hope so, as these are some of the very processes your brain uses to enable you to operate efficiently and convey these procedural implicit memories.

Priming is where we become influenced by unconscious information; that is, information that we are not consciously aware of. We will have been exposed to stimuli in the past that we may not be consciously aware of but they can (and will) still influence our behaviours now. An example of this might be when you find yourself singing a song, but you are unaware of why you are singing it. In reality, you may have been exposed to it earlier in the day but you did not consciously register it. You may not be able to recall the fact that you heard it on the radio or that it was a soundtrack to a TikTok that you watched. Subconsciously it was heard, though, and it is now influencing your behaviour as you are singing it. Another example might be if you go to the cinema and watch a scary movie. That will likely put you into a heightened unconscious state. This in turn increases the likelihood that when you come out (particularly if it is at night) you are going to be a little bit more alert, a little bit more on edge, maybe a little bit more 'jumpy' than you perhaps would otherwise have been. A final example relates more directly to marketing. If we see a brand at some point during the day, maybe on the side of a lorry or on a bag that someone has in the office, when we go shopping later in the evening, unconsciously, that exposure has the opportunity to influence the purchasing decision that we will make.

Both explicit and implicit memories such as these are what we would refer to as long-term memories. These endure, these last. It is very likely that we have had these memories for quite some time. Yes, we might be able to recall an event or a conversation that happened yesterday or potentially even earlier today, but we would still class that as being a long-term memory. And for a lot of what we do in marketing, long-term memories are really where we want to be focusing. The chances of us presenting content to somebody in exactly the right moment when they are considering purchasing is, for most of us, going to be slight. Therefore, what we want to do is build up awareness and recognition, so that when the moment comes and people are looking to procure a service or looking to invest in a product, they will think of us at that point. So that when it comes to making purchasing decisions, when it comes to exploring their options, to reaching out for quotes or looking online for reviews, ours will be an organization or brand that people are aware of and that may feel familiar to them. We need to be looking at these long-term memories, the different ways that they are formed and the different things we can do to support their formation.

In order to do this, I think it is also helpful to have a working understanding of the other types of memories.

Short-term memory

Short-term memories are usually only between 15 and 30 seconds long. Anything longer than that tips back over into what we would class as being long-term memories. So these memories can only be retained for a relatively short period of time, and they are finite in their capacity too. If I asked you to go to the local shop to buy the following 10 items

- butter
- lemonade
- dog food
- a pot noodle
- hand soap
- a birthday card
- fish fingers
- a lemon
- cheese slices
- tea bags

... how many of them do you think you could actually recall when you got to the shop? All 10? Maybe eight or nine? In most cases, you would only be able to recall between four and seven of these items... assuming you hadn't written them down or recorded a voice note on your phone, etc. That is the capacity of your short-term memory. It naturally doesn't enable you to hold onto lots of pieces of information. So, if we want to retain that information for longer periods, we are going to have to adopt some different methods.

Probably some of the easiest methods that we can adopt are using devices like rehearsal (where we simply repeat the items in list form), putting items in pairs, using visual methods so we activate the visual systems in our brains or even creating a tune or a song about it. All of these methods would help us to extend the amount of time we are able to recall something for and indeed the amount of things we are able to recall. There are lots of occasions when we rely on our short-term memory to give us those few brief moments where we need information for a few precious seconds. Just enough time to enable us to act on it or do something temporary with that information.

Sensory memory

Shorter still is sensory memory. This is a really interesting phenomena within the brain, where essentially a kind of legacy remains, even after physical stimulation has ended. It is almost as if there is a kind of trace that gets left when various parts of the brain are activated by the stimulation, as if it takes a few moments for that to then become deactivated. We really are talking about very short periods of time here – fractions of seconds up to perhaps one or two seconds at most. No longer than that. Unless we allocate resource to it because we decide that it is somehow meaningful to us. In which case, we may then invest the cognitive capacity in processing it and then it becomes more enduring.

We typically talk about three main aspects of this sensory memory (as shown in Figure 5.1) – iconic (visual), echoic (auditory) and haptic (tactile) – each of which correlates to one of our sensory systems through which the stimulations come.

FIGURE 5.1 Three main aspects of sensory memory

Iconic
This relates to the visual system. An example of this is when you look at a bright light and then when you close your eyes you still feel as though you can see that bright light for a few seconds afterwards. It is not as if you just shut your eyes and the view has gone. There is a trace element which is still left within your brain. You carry on experiencing it, even though that stimulation is no longer there.
Echoic
This relates to the auditory system. An example of this is when you are not really paying attention to somebody but you're aware that they are talking. However, if pushed, you could probably repeat the last few words of a sentence that they have just uttered. Numerous examples of this happen in classrooms across the globe I'm sure, where students effectively zone out, drift off, daydream. However, if a teacher suddenly calls me out and says 'Right, Katie, what was it I just said?', I could probably dredge something up from my echoic memory, a few words from their last sentence. So, again it gives us an enduring tail from that stimulation in auditory form.
Haptic
This relates to our sense of touch – our ability to recall the physical sensation of something. An example of this is continuing to experience the sensation of your mobile device vibrating in your pocket even after the actual vibrations have stopped. Or another common example of a haptic sensory is memory still feeling the sensation of touch on the surface of your skin once somebody takes their hand away.

In each of these cases, a sensual experience is created within us that lasts longer than the initial physical stimulation does. So, they are classified as memories that are being created, but they are created within the processing of that incoming sensory information.

Working memory

This form of memory has the shortest duration of all. It is a very pragmatic kind of memory, as it literally just enables us to complete tasks, and cognitive tasks in particular. It enables us to hold on to the small amounts of information that we need, in order to be able to complete some of the tasks that we do – all day, every day. These are the small amounts of information, for instance, that enable us to recall the start of a sentence as we read or hear it, so we can then understand the full sentence on completion. Other examples of working memory include holding part of a number in our head, while we complete a mental arithmetic calculation, or even holding onto the next step in a recipe that we are following, until the moment we carry it out. In each of these cases, we retain that information just long enough to carry out the task, and then it is gone. We don't hold on to that memory any longer than that.

All of these are examples of different sorts of memories we have, which can be defined according to their duration. And in their own way, they each provide interesting points about the ways our memories are formed. However, before we look at what we, as marketers, can do to support the formation of memories, there's one more, really powerful, really potent form of memory that we need to mention and understand.

Flashbulb memory

These memories are unique. They are often very detailed, very vivid and very specific. They will also often last for an incredibly long time... a lifetime in fact. These are some of the defining characteristics of flashbulb memories, the fact that we have huge amounts of detail and richness to those memories, which are way over and above any other memories formed at around that time.

Let me give you an example: *What were you doing on Monday 28 August 2000?* Any ideas? OK, I know, some of you weren't even born then. But for those of you who were, and who are old enough to recall it, do you know what you did on that day? What about the year after – Tuesday 28 August 2001? Now, unless I happen to have landed on your birthday, wedding day or other significant event, I would imagine that you are really struggling at the moment to be able to identify what you did on that day or even where you were (as in the UK these dates are within our school summer holidays). Now, what about if we jump ahead two weeks. What about Tuesday 11 September 2001? For many of us who were alive at the time, that date refers to a shared flashbulb memory – the terrorist attacks on the World Trade Centre and the Pentagon in America. If you ask about what we were doing the day before, it's very likely we've got hardly any memory at all. But, as soon as we talk about the day of the event, people can remember really detailed information. The classic aspects are that we can usually remember where we were, who we were with, what we were doing and quite often we can also remember solid amounts of detail for somewhere between two and four hours after hearing the news.

The key to flashbulb memories are that they are usually very emotionally charged events. It appears as though the level and intensity of the emotions experienced somehow changes the way the memories get encoded. It is important to note that the emotion could be very positive too, such as the birth of your first child, the evening you got proposed to, etc. Not all flashbulb memories are negative, and, as you can see, not all of them are global or even shared. Some of them can be very personal to us individually – either positively or negatively. The key is we need to know that we have these kinds of memories. We have some incredibly deep, incredibly detailed, incredibly powerful memories that get laid down very quickly and that last for a very long time. This is a fact that we need to be aware of and we need to be mindful of when we consider how we can utilize memory within our marketing.

The process of memory

One final consideration before we start exploring how we use these insights – the actual process of memory. We might hear people say 'Oh I've got a really bad memory' or 'He's got a phenomenal memory'; however, we need to acknowledge that when we study memory, there are three essential parts to it that we need to recognize and consider. If we want to be really effective in attempting to create memorable content, we have to try and support all three stages.

The first stage that we have is **encoding**. This is where information is taken from the environment around us and, through a process called long-term potentiation, it is encoded. The neurons involved have their synaptic connections strengthened, creating new circuits and the formation of a memory. This is then **stored** in stage two of the process. During the 1980s, research into our memories used the (then) modern analogy of a computer to illustrate the ability to encode information and store it and retrieve it. This felt like a fabulous parallel to the way our memory worked. Sadly, though, we soon learned that this is not accurate. Unlike in the case of computers, our memories are not static during the storage phase. No, they can be influenced, and very easily so. Therefore, when we consider the third of our three stages – **retrieval** – we need to appreciate that what we recover from our memories may differ from the original experience itself. Memories are not passively encoded and stored. They do not remain dormant until you want to retrieve them, and when you do retrieve them, they are not exactly as they were when they were laid down. The processes involved in creating, storing and accessing memories are active, and, as a result, our memories are alarmingly fallible.

Considerable research and effort has gone into understanding this phenomenon, and minimizing its impact in areas like the judicial system. Concepts such as 'leading questions', 'the power of suggestion' and 'false memories' may not have significant consequences in some aspects of our lives but, here, they can be vital. Numerous examples of research have shown that people's recollection of particular events and experiences can easily be influenced, distorted and even created, changing their fundamental recollection of the initial experience drastically.

As ethical marketers, I am sure this is not an aspect of memory that you particularly want or need to dwell on. So let us return to those areas that we will explore next – ways that we increase our chances of being noticed and recalled.

Memories and brands

Our idea of any given brand is actually made up from lots of memories, experiences and information that we come to associate with it. These are, in effect, memories that get amalgamated and filed under the schema of the brand name. Obviously, establishing brand awareness is all about creating these memories, relaying that information and adding on the relevant associations. This essentially is what builds up a brand in the first place. Without the ability to create memories, we wouldn't have any sense of brand. We wouldn't be able to differentiate one from another in a meaningful way. We could look at the practicalities, the very superficial differences between competing brands, but we wouldn't understand anything different about them in that deeper, richer way.

What can we do then in terms of enhancing our audiences' memory? Great question. So great in fact that it is a topic that we are going to come back to time and time again throughout the remainder of this book. Many of the recommendations I make and the components I introduce you to will relate back to this in one form or another. So, although it may feel as though I am going through this next section quite quickly, please, as my teenage daughter says, 'trust the process'! Once we get the basics in place here, we can then explore it in more detail in upcoming chapters, and look at the relevant aspects in much more depth. However, I hope you appreciate that you need to have this basic understanding and knowledge of memory in place first, before we can really get the true benefit out of the future content.

So, let us briefly list and summarize some of the most effective ways to enhance the chances of our brand being recalled.

Repetition, repetition, repetition

For many of us, the simplest way of being able to enhance our chance of being recalled is repetition. In marketing, many of us were told something along the lines of 'it takes seven exposures before somebody will remember you and before they will actually commit to a call to action'. Repetition is huge. Lots of advertising is based on repetition – just being there, just regularly being there. You essentially layer up the exposure someone has until it reaches a level where it gets consciously noticed. As we discovered when we looked into short-term memory earlier, rehearsal or repeating things is a good way of moving information over from your short-term into your more relevant long-term memory. We can use this to think about some of the strategies that we might implement within our marketing, effectively a suggestion to look at building on what already exists. We do not always need to generate new, massively exciting campaigns because we can actually achieve good results by building on what people already know and recognize. That allows the processes of rehearsal or repetition to take effect. People feel comfortable and feel familiar, instead of constantly being exposed to reinvented concepts and things being taken off in a different creative direction.

Rhyming

A variation of the repetition technique is to adopt the practice of using rhymes to reinforce the same concept. Instead of just repeating the same word or phrase, a rhyme can achieve an extension of this within the brain of the audience. However, the obvious advantage to turning it into a rhyme is that you can add more information, enhance the message and convey more details than simple repetition is able to deliver.

Emotion

Emotions are connected to memories in very compelling ways. If we can create emotional content, it gives us so much extra leverage in terms of being likely to build enduring memories. So we need to look to create content that isn't cold and distant but actually has feeling or, more accurately, that triggers feelings within the audience. Those then become, of course, the episodic explicit memories because they feel like they are events, like things we are experiencing and we are living through ourselves. It's advisable to try and introduce emotions wherever possible, to not only encourage our audience to connect with us in a more pertinent way but to be able to recall us more easily too.

Effort

This technique relates to the concept of cognitive effort. In essence, if we put increased levels of effort into something, we have an increased chance of being able to remember it. Putting that cognitive effort in requires us to apply attention – consciously or unconsciously. This need to become more involved in the content in some way is really going to improve our chance of recalling what we've seen or what we've been exposed to. This has so many possibilities for us in terms of the strategies and designs we create, each of which can subtly support people's ability to recall us and our key messages. However, this only works up to a certain point; if the content is too confusing, cluttered or appears to require an overwhelming amount of effort, our brain will choose to take the easy option, and simply move on.

Music

Music, and the effect that music has within our memory, is unparalleled. I am confident that you don't really need me to tell you that though. Have you ever had the experience of hearing the opening bars from a song you haven't heard in years, and yet, within fractions of a second, the words come back to you and you are there, singing along as if no time had passed at all? Maybe you can recall theme tunes to the TV programmes you used to watch when you were young or the jingles from adverts. These work so efficiently because of the way that music is received and processed within our brain, and particularly because of the unique relationship that exists between music and memory. Music holds great opportunities for use within marketing, and I think a number of organizations are beginning to tap into this concept of sonic branding in a really quite exciting way.

Smell

This may be harder for some of us to capture and benefit from, but smell is another element that has a unique position in terms of the brain and how it is processed. Smell triggers memories incredibly fast. Often before we can actually identify or articulate what a given memory is, we almost instantly recognize a smell and determine if it is good or bad. We may only have a sense of it at that early stage but, slowly, the relevant memory will be identified and a more detailed connection or experience can be articulated. The reason this happens is a result of the location and way that smell is processed within our brains – something we will go into in more detail in Chapter 13. For now, just accept that the recognition of smells happens very quickly, and their influence on our mood and behaviours can be very significant. Therefore, it is a consideration that many organizations are exploring and harnessing for brand leverage.

Primacy and recency

The final area I want to introduce you to at this stage draws again on the way that our memory processes work. If we want information to be remembered, we know that there are two stages when we stand a much better chance of that information being retained. They are at the start and again at the end of something. For this reason, it is referred to as the primacy/recency effect. We have already seen how the things that we pay attention to, the things we notice, stand a better chance of being recalled. This is the simple understanding that drives the primacy/recency effect.

We usually pay attention to things right at the start, as this enables our brains to make a decision regarding its relevance to us. However, once underway, our attention levels are prone to dropping off quickly. They do, however, return to being high again as we determine that the end of some content is imminent, which is in some ways similar to what we saw earlier regarding sensory memory. Therefore, right at the start and right at the end are where things are most powerful for us in terms of creating memories. These are the places where we want to really focus our resources, as they have the greatest level of engagement, which means they are also the most likely pieces of information that will be retained. These insights create numerous opportunities, across a variety of platforms, where we can harness and optimize the attention and recall levels of our audience.

Talking of content coming to an end, that concludes our brief exploration into the world of memories. We will be referring to these processes again once we go into more detail about the techniques and methods we can adopt as neuromarketers. Before we move on though, I have a question for you. How many of those 10 items on the shopping list I gave you can you recall now? No cheating now, don't turn back. How many of the 10 items can you name?

Chapter summary

1. Despite what we might say and think, we do not have just one sort of memory.
2. Memories can be divided into explicit and implicit categories, depending on whether we can declare and share them or not.
3. Short-term memories enable us to retain between four and seven 'chunks' or pieces of information, for less than a minute.
4. Sensory memories refer to when a legacy is left by our sensory stimulation. They can be iconic (visual), echoic (auditory) or haptic (tactile).

5. Flashbulb memories show us the critical role emotions play in our memory processes.
6. We need to acknowledge and address the three different stages of memory – encoding, storage and retrieval.
7. We have a number of devices we can use to support the way our brains process, understand and recall information about brands.
8. Adopting the primacy/recency effect is a powerful way for us to improve our chances of being noticed and remembered.

6

Defying the decision-making odds

Context

It is often stated that an average human adult makes in the region of 35,000 decisions each day.[1] Some of these will be unconscious and below the threshold of our awareness, whilst others we will definitely be very consciously aware of. Many of these will be small and apparently inconsequential, such as whether to have pasta for dinner or a stir fry. However, others will be large and have very clear consequences, such as whether to ask a loved one to marry you or hand your notice in at work. But what is happening within the brain when we make a decision, and what can we marketers do to facilitate the processes for our audiences?

The decisions that we make affect the lives we lead. That is profound, isn't it? Kind of obvious too! But think about it...

If you hadn't decided to go to university, you would never have met *that* person, who is still a major part of your life now. And think about the places you've been together, and the experiences you have shared. All as a result of that decision. Oh, and about a thousand others such as whether to answer their text, do you feel well enough to go out with them tonight or what to buy them for their birthday, etc. You get the general idea.

Decisions are a vital aspect of everyday lives. And as such, they take a lot of processing within our brain. Even the small ones, like deciding what to wear today or how to eat your Jaffa Cake, require us to weigh up some alternatives and then reach a conclusion that we will implement. But how does this actually occur within our brain? What are the mechanisms and processes that are involved in decision-making?

The area of your brain that is most commonly associated with decision-making is the pre-frontal cortex. This is the part behind your forehead, so right at the front of your brain. This region is implicated in many of our higher cognitive functions, such as planning, problem-solving, reasoning and, of course, decision making. I was taught to think of this area as the CEO within your brain. It co-ordinates the input from many other areas, and uses this information to create plans, solve problems and make decisions that are then implemented.

Consider an example: If I asked you to make a big decision right now, how would you go about it? If you had three options to choose between, where would you start? Shall we give this some context and see if that helps you at all? Suppose you have just won a competition. Your prize is that you can either:

a have an all-expenses-paid weekend break for two in Paris

b have an exclusive tour, for four people, around Harry Potter World after hours

c have Gordon Ramsay come round to your house to cook dinner for six people

How would you go about deciding? Maybe Gordon Ramsay is your favourite celebrity, so that could be a clear winner. Or maybe you have an aversion to Harry Potter since your child was in their Hogwarts phase a few years ago, so you can easily eliminate that one. How might you approach choosing?

Well, you could diligently write out a list of pros and cons for each of them, and add up the totals to see which has the most results. Or, maybe you could give each of the pros and cons a weighting and then add up the scores to see which has the highest rating. Yes, that might work... for a competition prize. But for choosing whether to have an apple or a banana with your lunch today? I don't think so.

And neither does your brain. Instead of diligently working through all the possible options available to us at any given decision point, it chooses a more efficient approach. It uses the benchmark of being 'good enough'. Good enough to not do us harm. Good enough for now. Good enough to probably be OK. Why would it settle for this though?

To put it simply, it is due to the limitations that we have within our cognitive processes such as memory, attention, etc. If we were to put all the required cognitive efforts into making a decision, we would need to use up expensive attentional resources, draw on an overwhelming amount of previous memories and carry out some phenomenal mental arithmetic (holding each of the options in our mind at the same time remember) before coming to our conclusion. And how often would that be an effective use of our processing power? Instead, we adopt the 'good enough' attitude, which allows us to quickly make a choice, and then move on with the rest of our day.

Do not misunderstand me though, it does not just guess. It does attempt to evaluate the options, using a combination of data available to it and any relevant information that may be stored within our long-term memory. Have you been faced with this choice before? What happened last time? Did it work out well for you or were there negative consequences? Maybe you have read or heard something that may be relevant to the choice you are presented with. All of which sounds very thorough and helpful. However, there is a missing ingredient that we need to consider here. Can you guess what it is?

Yes, you are right (you see how much faith I have in you!).

Emotions

Much as we may like to think that the decisions we make are rational and follow logical drivers, they are in fact driven more by our emotions. Do you remember in Chapter 4 when we looked at the Ultimatum Game? Can you recall how easily our rational choices get overridden by an emotional desire to punish someone or to penalize them for the unfairness of their offer?

This emotional activity within the limbic system of your brain is also part of the information received by the prefrontal CEO. But it must be noted that this information may often lead us to biased responses and inaccurate perceptions, as it is based on a selective and skewed foundation.

Take this well-known example that pops up on social media from time to time. Would you rather have:

a £1 million now, or

b 1p today, but it doubles every day for the next 30 days?

As we try to comprehend the rational argument of which is likely to be the best choice for us, our emotions get right in the way and throw doubt on things, and attempt to secure certainty against risk etc. Just for the record, the best answer would be to choose option b. Despite what we may think about how this would pan out, if you were given 1p today and it doubled every day for the next 30 days, by the end of that time, you would actually have £5,368,709.12. If you don't believe me, go and check it out for yourself.

That should've been a relatively easy choice, and one where our emotions are unlikely to run too wild (with the obvious exception of the fact that someone has just offered you £1 million!). Can you imagine how much more influence our emotions have when you are deciding whether to take Job A (which pays more but means I have to work away from my family two days a week) or Job B (which pays less, but means I will be able to see my children each evening, and take them to their clubs, etc)? So much emotional content to weigh up and consider there, and your final decision will say a lot about your core values, motivations and priorities.

Another neuroscience element that we need to consider in reference to decision-making is the influence of neurotransmitters. In Chapter 2 we learned about dopamine, which is produced when you anticipate a pleasant outcome, a good result or even just positive feedback. This is clearly going to be important to the decisions you make then isn't it? You want to choose the option that is going to be the most pleasant and rewarding. But remember how powerful dopamine can be. It can lead us to repeat behaviours, with long-term negative consequences just because they provide short-term positive ones. Think about addictions, and how hard it is to break out of a cycle of using alcohol, drugs, gambling, etc even when we know that they are causing us harm. This is the power that dopamine has over the decisions you make.

Similarly, the levels of serotonin within your brain can dramatically affect the decisions you make. It is critically involved in the analysing of risk vs rewards. It has a correlation with our ability to control or inhibit our impulses. And it is also implicated in our levels of motivation for completing a task (as high levels of serotonin reduces the amount of effort we believe will be required in order to complete it). So, the same choice, given to people with low and high levels of serotonin can lead to very different decisions and outcomes.

We have learned that these elements within our brain can affect the decisions we make. But what else can?

Internal elements

Emotions

Let's start by going back to the area of emotions. We know that emotions and the limbic system have a significant influence on the decisions we make. But what is the nature of those? Well, the honest answer is that there is no straight-forward answer to that question. You see, much depends on the emotion you are experiencing.

If you are feeling confident, enthusiastic and generally buoyant, then you are likely to make riskier decisions. When we feel positive and things are going well, we are more inclined to take chances, to view things more favourably and make more audacious choices. Conversely, if we are feeling down, or are anxious and fearful, then we will do the opposite. Experiencing negative emotions means we are likely to underestimate the chance of things working out well, and so we take more cautious decisions, seeking greater certainty and security.

These influences are particularly apparent when it comes to making decisions that have a high degree of uncertainty or ambiguity in them. Just to be clear, within the study of decision making, there is a significant difference between decisions that are uncertain and those that are ambiguous. They can be summarized as follows:

Uncertain – this refers to decisions where there is a number of possible outcomes, and we are uncertain about which of these is the most likely. It is usually focused on future scenarios, and we do not have enough information or knowledge to enable us to decide with certainty. For example, we may be uncertain about what to wear on Saturday because we do not have enough knowledge about what the weather is going to be doing then.

Ambiguous – this is different. Ambiguous decisions occur when there is not enough clarity or instruction. The situation does not provide enough information, so it is up to individual interpretation to decide what was intended or requested. Ambiguity is usually focused on the present, and is experienced when the options are vague and unspecific. The instruction 'buy dinner' is ambiguous – how many people for, are they fussy eaters, do we have any allergy needs or preferences, maybe you are just paying for dinner not actually sourcing it too, etc?

Interestingly, research has shown that the processes involved in decision-making within our brains are not the same for both types of decisions.[2] Although some areas do overlap and are clearly involved in both types of decisions (such as the anterior insula – part of the cerebral cortex), others differed between the two types. This shows that our brains approach the two situations differently, and achieve their conclusions as a result of accessing fundamentally different information.

Personality

Building on from our emotional state, our personality clearly has an influence on the decisions that we make. If you are someone who likes to seek out new experiences, wants adventure and is generally comfortable with uncertainty, then you will be more inclined to take more chances and higher risks. Whereas, someone who is more cautious, more risk averse and who likes predictability is more likely to make conservative, safer choices.

Biases

Even though we may prefer to believe that we are above being biased, we actually all exhibit several forms of bias on a very regular basis. For example, there is a phenomena called confirmation bias, which occurs when we have a core belief or set of values. Maybe you believe that men and women are fundamentally equal – different, but equal. Now, this belief means that your Reticular Activating System (RAS; see Chapter 3 if you need a reminder about what this is) will naturally select and draw your attention to information, examples, evidence and experiences that corroborate this pre-existing belief. Conversely, it will also 'play down' anything that does not support the view you hold, and even if you do notice it, you will give it less credence than something that substantiates that view. Therefore, the biases we hold can affect not only the information we see but also the value we place on it.

Motivations

The final internal piece of the jigsaw relates to our motivations around the decision. How many purchases are made each year, motivated by the way we will be perceived? Owning *that* car, wearing *those* clothes or being the first to try *that* new app all reflect on us as a person. So is that something that motivates you when you make a choice? Maybe you are not motivated by that so much as the impact your purchase will have on the environment or the negative consequences that will be experienced if you buy the more expensive one. The way we prioritize our motivations will help us to ascertain which of the many pros and cons being considered are the ones that hold the most weight for us. However, it is vital that we remember that much of this will be happening at a subconscious level. So what we might consciously think of as being our priority (save our money for going on holiday in the summer) can often be overruled by our unconscious priorities (not to disappoint our children and let them down). So, we end up buying 'the thing' they have been going on about incessantly for ages and ages...

Now that we have an awareness of the internal factors that affect the decisions we make, let us shift our view and consider the external ones. After all, these may be where we as marketers can have the most influence.

External elements

Time pressure

When you are required to make a decision rapidly and under time pressure, you are less likely to make good choices. The urgency creates a stress response within you, and this in turn reduces your ability to process the options effectively.

Many marketers attempt to harness this element by promoting a sense of scarcity around their products or services. 'Only four left at this price' or 'Your exclusive offer ends in 3 hours, 11 mins and 8 seconds' or 'This year's challenge starts on Monday morning at 9:00am'. Can you see how they are all creating a pressure in you, to make your decision quickly?

However, we do need to be careful about how we deploy techniques such as this. It has been shown that, yes, using them can increase sales, conversions and even the number of people you get signing up to your challenge, but proceed with caution. People who make purchasing decisions in haste are much more likely to regret them in due course, thus creating a highly significant increase in returns/complaints of 17 per cent.[3] The cost of servicing those could easily make it a much less efficient marketing strategy to adopt.

Peer pressure

Another external element that can significantly affect your decision-making processes is the presence, profile and behaviour of those around you. Knowing what decision other people have made often changes the decision we make. We may seek to 'fit in' and so decide to prioritize that over any other factors that were previously influencing our choice. Or we may be more motivated by not conforming with that particular group, and so choose to do something different in order to confirm our status that way.

Similarly, if the individuals around us are perceived as being authorities in the field, then we may place greater weight on their choices and recommendations, as we believe them to know more than we do. It is worth noting here that our ability to determine who is an authority is very easily manipulated and influenced. A reported job title, a claimed connection or association, even just the clothes people wear, will establish a sense of their level of authority within our minds. And these perceptions will go on to have profound effects on the decisions we make.

This is a relatively easy area for us to harness the potential of within our marketing strategies and campaigns, and is generally encapsulated within the idea of social proof.

At its most basic level, the use of reviews from organizations such as Trustpilot or Google can provide a tangible sense of reassurance about the purchase we are making. Have you ever stopped to consider how much weight you place on the opinions and experiences of people who you have never met? If 1,482 people have reviewed it, and give it an average of 4.4 then it must be OK... right? But what if only eight people have reviewed it and given it an average of 4.4? Would you have different feelings about your purchase now? In the absence of having reliable evidence to support our decision-making processes, we fall back on numbers. High numbers tell us that lots of people have made the same choice as we are considering, and there can be comfort in that for our overwhelmed ancient brains. The likely risk is going to be perceived as being much lower, if we are doing what others have already done ahead of us (and, critically, have not apparently regretted their choice).

Building on this is the importance of having testimonials, case studies and endorsements for our services or products. This not only shows that others have made the same decision, it tells us more of the story behind their situation and circumstances. The impact of these goes to another level, if they can be provided by a person or organization who is well known and well regarded. This could be because they are an authority within the field, a household brand or simply considered a celebrity. All of these will skew the value we place on their comments, value that is already higher than anything we might say about ourselves because, let's face it, we are going to say good things about what we do aren't we? But when someone else says good things about us, it is so much more believable and therefore powerful. Hence its potential to influence the decisions that we make.

Easy decisions

One of the most common areas where neuromarketing produces results that are the polar opposite of what customers tell us is regarding the need to have choice. In the vast majority of research we carry out, most marketers will regularly be told by their audience that they want more choice.

'What do we want? Options. When do we want them? Whenever we like!'

Interestingly, the brain takes a very different view. When we use fMRI (functional Magnetic Resonance Imaging) to observe the brain making decisions, we can see that it very easily becomes overwhelmed, and is paralysed by the number of options available to it. And so it seeks the 'good-enough' option, which very often is to continue with things as they are, i.e. we choose to not make a decision. Or, the alternative, as we have already seen above, is that we give too much power to things that are not necessarily relevant to the decision we are attempting to make – such as the reviews for it or even just the price.

One way that we can support our customers and facilitate their decision-making processes is to help their brains to easily identify what the best option is going to be. How do we do that? By allowing them to easily make the necessary comparisons. Imagine that you are keen to purchase a training course for some of your team. When you identify a possible provider, you are given the following list of options:

a one full day of training, delivered off site for your whole team

b two half-day sessions, delivered online and for a maximum of 12 delegates, or

c download the modules for your team to study independently at a time of their choosing

How easy would you find it to compare those options? Even before the critical concept of pricing becomes introduced, these are hard to choose between because we are comparing very different options. Essentially, there are too many variables. A more brain-friendly approach to take would be to allow them to select from fewer options each time:

- **a** Would you like your training delivered in person or virtually?
- **b** Can you train the whole team at once or do you need to do it in smaller groups?
- **c** Would you prefer one full day of training or to split the training, allowing time in between for application of the content?
- **d** Do you want the content to be taught collectively or could individuals access it at a time of their choosing?

Can you see what that does? It allows the customer to choose their own package, but one question at a time. Our brain should be able to deal with a choice between two options at each stage!

Before we move on from the idea of supporting customers with their decision-making, I would like to take a moment to draw your attention to two other areas that are relevant. These could be considered as more 'sales' elements than marketing ones, but I feel like I would be letting you down if I did not make you aware of them here.

Objections

We have to accept that many of our prospects will have some barriers that prevent them from instantly signing up, purchasing, subscribing, investing, committing, etc. These may only be minor or they could potentially be very significant. Either way, we need to address them. I appreciate that this may take us into areas we do not want to visit, but we need to be open and honest about these concerns, and so the best way for us is to face them head on.

The first question we need to ask ourselves is what we know about the nature and form of these objections. Do our sales team note the comments that are made or does anyone monitor the chatbot for trends in the questions asked. There are many possible sources of information that could give us ideas about the objections our prospects face. Identifying these and bringing them into the light does not make you any less successful as a business. In fact quite the opposite. Raising these and addressing them will actually give you a series of significant advantages.

Being honest and transparent

When a prospect encounters an organization who is openly addressing concerns that may be raised, they will associate the organization with being transparent, honest and ethical. Instead of trying to deny there may be any concerns or queries, an organization that openly raises and addresses them appears to have nothing to fear or to hide.

Being seen and understood

If the nature of the objections identified are an accurate reflection of the questions, concerns and considerations that the prospect has, they will feel very seen and understood. It will show them that you understand their situation, you are aware of what they may be feeling, and you are happy to support them with any information they need in order to make their decision.

The precise way to deal with these questions will vary according to the platforms and formats that are being used. Clearly, the ideal scenario is if these can be uncovered and addressed in a conversation, either in person or over the phone. However, some highly effective use can be made of FAQs and myth-busting social media posts, which outline the question or concern, and then provide a full and frank answer. Although this may be more generalized and need a little more digging to uncover, it does also mean that the information will be available for a large number of people to access and appreciate.

Creating content and connection

Utilizing this approach can not only build confidence and trust but it can also help significantly with SEO and AI solutions too. Writing blogs around the questions that you have identified or videoing members of the team responding to them will again give you lots of opportunities to raise your profile and reputation. However, remember to have a visible call to action (CTA) at the end of all of these pieces of content, because if you have managed to address the concern they had, you want to let them act on that right away. Which conveniently leads me to this next point.

Closing

The final part of this process to consider is the idea of closing the sale/deal. No-one likes to feel that they are being put under undue pressure to make a decision, but nor should we just leave the whole thing to just fade out. We therefore seek to get a balance between the two.

The key to doing this successfully is to allow the prospect to feel like they have control. Ask them if they have got any questions, clarify that they have got all the information that they need and get them to suggest what they would like the next step to be. Remembering how difficult our brains find it to make a decision, we may need to help them with this though, and if that is the case, ask them if it acceptable for us to make a suggestion. There are two concepts that may transform your success rate when it comes to closing.

1. Presumptive close

We know that the brain is going to struggle if we leave it with too many options and choices to consider, and the result of this may invariably be that no decision gets made. So, we can help the prospect by providing them with a small number of options to choose between... ideally two. Instead of leaving them to decide the timing and steps from here, consider saying to them 'I appreciate that you may need time to consider what we have suggested today, would you like me to call you next week, or the week after?' Can you see why this is called the presumptive close? You assume that the deal is happening, and in most cases their brain will go along with that assumption too. Clearly, this needs to be done in ethical and sensitive ways that honour your brand, tone of voice, etc, but hopefully you can see the concept I am suggesting.

2. *Law of consistency*

One other element I want you to be aware of before you go implementing some of these ideas is referred to as the law of consistency. Now, this term is applied to a number of different scenarios, but in this – essentially sales – case, it refers to the way we can start small with a client and build their involvement up. Let me share a piece of research with you, which demonstrates the potential of this. Residents in a neighbourhood in America were asked if they would be prepared to have a 6 ft × 8 ft board placed in their front garden promoting safe driving. If they accepted the offer, they would be paid $100 per calendar month (pcm) for their involvement. Unsurprisingly, only 1 per cent of people took this offer up.

However, in a comparable neighbourhood, just a few miles away, the results were drastically different. In that instance, 7.5 per cent of people accepted the offer. Why? Because in the second neighbourhood some of these residents had already been exhibiting a postcard in their front window featuring the same design. You see, when they refused the offer of the board, they were then given the chance to display a postcard for $10 pcm. Here, 30 per cent of people accepted the postcard offer first time round, which was a good result, but clearly would not make for great visibility due to the size of the content being displayed. But, having displayed these for a short period of time, they were then approached about upgrading to the full board, and 25 per cent of them accepted, making a 7.5 per cent conversion overall.[4]

This shift is thought to have been as a result of them now having a connection with the organization, knowing that they can trust them, having some existing trading relationship in place and realizing that no adverse effects had been experienced. Whatever the precise reason, the key takeaway from this is that if someone is not ready to invest, subscribe, commit *yet*, then it would be good for us to have something smaller and more accessible that they can engage in at entry level. So, what could you offer? Do you have a sample module people could buy, a seven-day plan they could sign up for, a taster session they could attend, a smaller product they could purchase or an initial, no obligation, consultation they could access? Holding this back until they have made their decision regarding the initial, larger, offer will support their decision-making processes and allow you to present it as a more 'exclusive' element that will be valued more highly.

B2B vs B2C

Before we leave the topic of decision-making, we need to address one major elephant in the room. That is, the belief that these concepts only work in B2C.

Please. Have I taught you nothing so far?

If you recall, the part of our brain that determines and drives most of our thoughts, actions and behaviours is just focused in the present. It is not able to contemplate the future or the past, it is non-verbal and it has a very limited range of priorities that it is looking out for. Do you really think this part of our brain is going to respond differently to content and materials depending on whether it is in 'work mode' or 'home mode'? No, it is definitely not. So, many of these influences occur in B2B decision-making too.

However, let's be honest here. Very often with B2B decisions, we try to convince ourselves that we are making grown-up, rational decisions that are well thought out and not based on biases, emotions, etc. And how do we do that? By getting lots of people involved. That way, we feel as though we stand a better chance of making the 'right' decision. In practice, of course, this simply means that we have a number of individuals' personalities, emotions, biases and motivations to now work with instead. To say nothing of the influences that group dynamics and company politics can have on the overall process.

Pardon my tone there, but I hope you can sense the passion that I am writing with. We need to be thinking about marketing human to human, regardless of whether that human is at work or not. Indeed, the technology has enabled us to blur the boundaries of work and home lives so much that few of us have much demarcation between it anyway. The essential part of our audiences' brain that we need to consider and target is not going to change anything in the way it processes information from the world around it. Therefore, we have just as many opportunities to gain from these insights in B2B as we do in B2C. Arguably more, in fact, as few people within B2B marketing have truly harnessed the power that neuromarketing provides...yet.

Chapter summary

1. We make thousands of decisions each day, many of which we aren't even aware of.
2. Decision-making is hard for our brains, and involves the consumption of expensive cognitive resources.
3. Our brain works on the basis of 'good enough' to save it from decision paralysis.
4. The neurotransmitters dopamine and serotonin will significantly affect the decisions that we make.
5. Other factors affecting our decisions may be internal, e.g. our emotions, personality, biases and motivations.
6. External factors that influence our decision-making include pressures we experience as a result of time or our peers.
7. We need to support the brains of our audiences, and provide them with easy decisions to make as a result of simple and clear comparisons.
8. These elements are as significant in both B2C and B2B, as the most influential aspects of our brain do not change according to whether we are at work or at home.

Notes

1 Original source unknown, but this figure is cited in numerous publications such as the *Harvard Business Review* (https://hbr.org/2023/12/a-simple-way-to-make-better-decisions (archived at https://perma.cc/86TP-GHS9)) and *Psychology Today* (www.psychologytoday.com/gb/blog/stretching-theory/201809/how-many-decisions-do-we-make-each-day (archived at https://perma.cc/SG2C-32VA))

2 Wu, S, Sun, S, Camilleri, J A, Eickhoff, S B and Yu, B (2021) Better the devil you know than the devil you don't: Neural processing of risk and ambiguity, *NeuroImage*, 236, https://doi.org/10.1016/j.neuroimage.2021.118109 (archived at https://perma.cc/DP5Y-4YDN)

3 Calvo, E, Lui, R and Wagner L (2020) Hurry! Only 3 left in stock! When scarcity signals are most powerful (28 April), IESE Insight, www.iese.edu/insight/articles/online-retail-scarcity-signals/ (archived at https://perma.cc/CZG5-RF79)

4 Cialdini, R B (2007) *Influence: The Psychology of Persuasion*, Collins, New York

PART TWO

Putting theory into practice

7

Getting to know a whole new you

Context

In order to truly benefit from most of the insights I am going to introduce to you over the coming pages, you are going to need to get really clear about who your audience is. I am sure you already feel that you have this knowledge, that you know them well enough and that we should just get on with the good stuff. Well, patience my friend, because I am going to encourage you to try and approach this challenge in a different way.

As professional marketers, we are used to researching our audiences to find out what we can about them. We then use these insights to create segments, dividing the target population up according to shared traits and characteristics. Yes, as much as we all hate the thought of being put into boxes, that is what we have to do in order to create relevant and engaging products, services, content and campaigns.

However, as I am sure you are beginning to appreciate, the application of neuroscience brings with it the opportunity for you to see and do things differently. The difficulty is that it also brings challenges too – if we cannot rely on what our audience overtly tells us, how are we expected to be able to know them at any greater depth?

First, let me reiterate what I said in Chapter 2. I am not expecting you to be able to rush out and commission a full-scale piece of neuromarketing research on your prospects and customers, in order to gain the awareness and insights you seek. Sadly, that is still beyond the reach of many budgets and practicalities. However, do not lose faith!

There are things you can still do, and we will explore them below. So sit up, take notes and be prepared to revisit some of what you think you know about your audiences.

The challenge

If we ask our audience about their preferences, desires and pet hates, what we essentially get is information relating to their wants and needs. And that is fine. Provided, however, we accept that it is not going to give us any great insights into what we need to understand, emphasize or change if we are to truly connect with and serve this group. And, in case you have been reading this book in your sleep so far, this is because they are not aware of their true motivations and feelings themselves, so they are completely incapable of reporting these reliably to us. Therefore, we need to move beyond their wants and needs, and dig deeper.

Below the superficial, consciously accessible information we provide is the good stuff. Not only is this the source of what really drives our future decisions and behaviours, but it is also much more accurate than what we consciously know. Very often when I conduct research, I end up with at least two different sets of data: those provided by what people say, and those provided by the physiological measures I gather. These different data sets can present very different information but, time and time again, it is the physiological ones that turn out to be the true indicators of future behaviour. It takes courage, but these are the sources we need to analyse, interrogate, understand and base our decisions on, not what we are told.

A great example of this is the work I did for a B2B client back in 2020. There were essentially three elements to the research that we did:

- **Overt responses** – gathered using a traditional survey method
- **Physiological responses** – gathered using EEG
- **The real-world experience** – gathered using split testing

After conducting many hours of research with a wide variety of participants, we started to process the responses. Two very different pictures emerged: what their mouths were telling us and what their brains were telling us! Thankfully, in this instance, we also had the results of the split tests too. And these totally aligned with the physiological responses. The EEG responses had indicated strong preferences and aversions for specific content, and we could see that the inclusion of these design elements lead to dramatic changes in engagement levels, view times and conversion rates, when the designs appeared in real-world scenarios.

Of the three sets of information, sourced from the surveys, EEG and split tests, two out of the three aligned well. The anomaly was the surveys. What people told us indicated responses that differed significantly from their actual behaviours. The way people behaved and responded to the materials in context differed greatly from the way they told us they would behave. The only way we can accurately gain insights into what our audience is likely to do when products are launched, materials are produced and campaigns are released, is to ignore what we are told and focus on physiological measures instead. These cannot be skewed, they cannot be controlled and they are not susceptible to biasing. The raw, reliable information you need lies here.

I know, I know. I stated that this was not going to be about you commissioning large pieces of research, and it genuinely is not. So, now that we have reminded ourselves of the issues we face, and the need to do things differently, let's explore some ways that we can all begin to actually achieve this, in practical terms.

Ask your customers

Pardon me, what? Did you just read that right? After everything I have just said, I am suggesting you ask your customers? Yes, I certainly am. Let me explain why.

Interviewing or surveying your existing customer base will give you some gems of insight. Admittedly, it will not tell you everything you need to know, but it will give you some valuable information. So,

I am not saying don't ever talk to your audience or customer base again. I am saying that it has its place, and provided you are clear about its limitations, it has some validity.

For example, consider asking your customers and key accounts to share information such as:

- Describe the situation they faced prior to working with you. What were their concerns challenges, fears and risks?
- Detail a specific situation where they felt out of control or frustrated because they did not have your product/service.
- Can they quantify what was at stake? How much money, time, market share, reputation was on the line if they had not decided to work with you?
- Specifically, how did working with you help provide stability, create certainty or reduce the level of risk they were exposed to?
- Describe the difference that your products/services have made. What impact have they had? How have things changed for that person individually or their family/organization?

The key factor that all of these have in common is that they are getting people to articulate things that they can report – facts and situations. These will typically be much more accurately presented than the less tangible areas of emotions and motivations. However, there are other reasons why I am suggesting you ask them these questions.

Stop, look and listen

If you are able to do so, record the interviews that you conduct. Whether by video, through online meeting platforms or even just using audio, recording their responses will be invaluable. First, as I am sure your research experience has already taught you, recording the responses in unobtrusive ways means you are more likely to build a rapport, get the discussion flowing and generally connect more effectively with them. If you have to keep pausing to log or write down their responses, it breaks the natural rhythm of human conversations, reminds them that they are part of a process and it also breaks off important cues such as eye contact etc. Whether they are prospects or existing customers, removing barriers to them being able to converse with you openly will improve the quality of the information you receive.

The second reason to suggest you record the conversations is so that you can slow them down and analyse them for any cues you get via facial coding or voice changes (see Chapter 2). You may not be trained or qualified in these areas, but if you are prepared to put a small amount of effort in and, critically, to trust your instincts here, you can pick up on some points that will help you begin to delve deeper than what they are telling you.

The third reason is I want you to start really tuning in to the way they speak. Their words, their phrases and their descriptions are all powerful material for you to use in future communications. Think about it. How often do we use 'corporate' jargon and language to describe what we do? Phrases like 'provision of bespoke solutions, tailored for your unique situation' are all too common in corporate communications. But who talks like that? How much easier is it for our brains if we can use language that resonates with us straight away, without us having to essentially translate it?

Capturing the way they describe their situation prior to working with you will give you gems you can use to attract attention and create connections. Consider the following example.

Current text:

Bookkeeping solutions that efficiently manage all your accounting needs.

Customer phrases:

'I just never felt like I was quite on top of our accounts.'

'I felt trapped by the constant need to find time to do the accounts, and terrified that if I got it wrong, I'd end up in court.'

'I was always playing catch-up instead of feeling like I was driving the business forward.'

'It wasn't my expertise. I just wanted to get on with doing what I do best.'

Improved text:

Bookkeeping support that releases you to do what you do best.

Can you see the difference? More importantly, can you feel it? The 'current text' feels distant and cold. Clinical even. Whereas the 'improved text' creates more connection. It is something the audience can lean into and have feelings towards. It also uses powerful words like 'releases' which shows an understanding of the sense of fear they may experience, and the barriers, limits and mental weight that these onerous duties often produce.

Using their kind of phrases and language will help your prospects to feel seen and understood. It will encourage them to bond with your organization more easily and trust you more quickly. Critically, it will also use less cognitive resource in the process of doing all of this too, which will leave them feeling less overwhelmed and more empowered to make a decision.

Put yourself in their shoes

This is a process that many of us have tried to do successfully over the years I am sure. However, this time you have some neuroscience to add into the equation, so I am confident that you will get more from it than you have managed to achieve before. As I have told you my daughter would say, just 'trust the process' and see what you discover!

The way to do this is to take one segment or persona at a time. Now, based on all your experience and everything you have learned about them, try to put yourself into their shoes. What is it that they want or need? (#1)

Great. Now, don't just accept that at face value. Remember, we need to dig deeper. So, why do they want or need this? What will it do for them or provide them with, or allow them to do? (#2)

Good. Now, deeper again. Why do they need this thing that they will get or be enabled to do? (#3) Do you see what I mean? No? Then let me give you an example.

Say we are in the business of marketing drone videography.

So, our initial response to what one of our customers might want or need (#1) is footage of their hospitality facility to showcase its location within acres of parklands. Now, why do they want that?

Well, we might assume this need (#2) is based around them wanting to differentiate themselves from the competition, demonstrate their USP, etc. Fine. But why are they looking to do that?

Maybe their justification for this (#3) is that they have recently experienced a decline in enquiries, which they believe might be due to new entrants into the sector who are located closer to the main road network.

Now we are getting somewhere. But can we go deeper still and into level #4? Maybe we believe this concern might be as a result of still not seeing a return to pre-lockdown figures, and being concerned that the owners may decide to close or sell as an ongoing concern.

Now do you see what I mean? If we keep digging deeper, we will get below what people may have told us, and we can start to anticipate what they might really be feeling underneath. Maybe they have thought this at some point or maybe they have not. Either way, these are more likely to be the motivating factors that we need to tap into.

If we delve down deep enough, we usually find we end up with some very core, base fears or pains which the reptilian part of our brain is motivated to address. Hence why they influence our behaviours and actions so well. These can usually be put into a small number of categories.

Security

This is rooted in fears based around the resources that we need to keep ourselves, and what is important to us, safe. These relate to our basic drives of providing for ourselves, and making sure we have sufficient resources to cover our immediate, and anticipated, needs. Clearly, they originate from our ability to secure and maintain resources such as adequate food and tools, although more recently this is often reflected more in financial terms. This includes motivating factors such as the fear of not having enough and, more intensely, the fear of losing what we already have. It appears as though the risk of losing what we already have is roughly two times more motivating to us than the possibility of receiving greater volumes.[1] You will learn more about this is Chapter 12, but for now, just accept that this 'loss aversion' contains a significant amount of potential for us in terms of understanding our audiences.

Status

These fears are based around our position and how we stand in the eyes of others. These are similar to the security category but they are not around physical resources such as money, food, etc. These are based more on our reputation and our ability to maintain our position within a group. Fearing that we do not have enough knowledge, information, data, etc falls into this category, as does the fear of not having enough influence and control over our situation or circumstances.

Strength

This final category of fears is based around our personal power and ability to act. These relate much more to our sense of self and how we feel about who we are. Status is how others view us, but this is how we view ourselves. Do we feel we are worthy or capable enough? Are we too stressed or anxious, and do we feel safe in our continued survival? Essentially, these fears are a reflection of our own sense of the power, competence and confidence in ourselves. Are we likely to survive this? Because the ultimate fear that our limbic system has to mitigate for is our threat of extinction. Game over!

B2B marketing considerations

Now, you might be thinking that this is all well and good for those of us who are in B2C, but how does this relate to B2B marketing? Good observation. However, the response is that precisely the same core concepts apply. We may need to apply them slightly differently, but the same concepts will be driving our behaviours at work as when we are at home. So let's see how that might look…

Security

Here, again, it is about our resources and our ability to provide for ourselves and those who are important to us. So it may be about cashflow, having sufficient profit margins and ROI, having reserves and the tools, equipment, etc, which is required to give us confidence about our future.

Status

This one is fairly easy to translate. This is about our brand, positioning, profile, reputation and market share. It is also about knowledge too, remember, and the information we have access to, the data we monitor and the control or influence we feel we have.

Strength

What is our culture like? How do people feel about working for us and being part of what we are trying to achieve? Do they have faith in the leadership team, and do they share the corporate values and buy into the corporate objectives?

Now, underneath all of these company issues and challenges, we have to accept that the reptilian brain of your audience will also be working for them individually – we have learned how self-interested it is already haven't we? Therefore, we may have additional layers underneath these, which bring us back to the individual and what it means for them personally. For example, if an organization is struggling financially and margins are being challenged, it is not hard to see that this could also trigger individual fears around security and our ongoing ability to provide for ourselves and those around us.

The key message here, to mis-quote a mash-up of *Strictly Come Dancing*'s catchphrase and a World War II poster is... 'Keeeeep Digging'! Just keep challenging every level you reach and see if you believe there is anything further you can discover or anticipate underneath it.

Now, because you are making assumptions here and trying your best to understand what the situation *might* be for your audience segment, we have to accept that this is not going to create a perfect outcome. What you will do, though, is develop a much better awareness and understanding of what *could* be affecting them. Then, you can use these insights during your conversations and communications to see what responses you elicit. Over time, you will evolve this into more accurate knowledge, as a result of reviewing and refining the responses you achieve.

So, now we have an understanding of what might *really* be motivating your audience, the next stage is for us to explore ways to demonstrate that we are able to resolve some of these for them. There is little point in us just showing that we understand their situation unless we follow that up by sharing that we can help them to improve it. We need to not just sit with them in their deepest fears, but give them a way to reduce them or better still, alleviate them and their effects.

The quick way to do this is to just 'flip' the fears we have identified, and present them in reverse form as a solution. So, if we understand that a primary fear might be the lack of control they have over their staff absence rate, then we can create messaging based around solving that for them. This is what I did in the example above when we were talking about using their language. I didn't just mirror their words back to them, I presented them with a solution for it.

Finalizing the customer's decision

Now, there are some critical elements to this that we need to address. Although we may be able to appreciate that using phrases that are familiar to us, and being given a pathway to addressing our base fears, is going to be appealing to our limbic system, we still have work to do. Much as you might like to think that this is pretty much 'job done' and the prospects should now flock to us, this is not yet the case. You see, we have still given their brain a lot of work to do. And knowing how overwhelmed it is, how it likes to take the easy option and how it struggles with making decisions, we can expect that these three factors will combine to mean it still might not conclude that action is yet required. Our audience may be curious, interested even, but still there are barriers to them making a purchase or some other form of commitment. Why is this?

The likely issue is that we have still left too much work for their brain to complete. We have not made it truly compelling for them to reach the desired conclusion we want. This work is required in order to really understand that time old marketing question… 'what is in it for me?'.

The key to helping your audience understand the answer to this, and do so without having to put much effort in, is for you to put the effort in instead. And this all comes down to two things:

1. making it tangible
2. the burden of proof

Let's examine these in order, so you can build a really easy, frictionless route to take your prospects through to conversion.

1. Making it tangible

Here, we want to present them with the most complete version of what they can expect – with all the vagaries and fluffiness removed. So, consider the following example propositions and see which one you have the most confidence in:

- **a** we can help you reduce instances of staff absenteeism
- **b** we can help you reduce instances of staff absenteeism by 12%
- **c** we can save you an average of £2,872 per month, by reducing instances of staff absenteeism

Many of us use phrases such as that seen in the first example in order to promote the benefits of what we provide. However, if we do not quantify these benefits, we are missing a major trick. Let's examine why.

In the first example the word 'reduce' is good, but it leaves a huge amount of interpretation to be done. One person may believe that it really suggests a small reduction, say 2–5 per cent. Another could assume we mean we can half their instances. Whatever their natural level of expectation and experience is will determine what they factor into the decision that they make. So someone who is naturally positive and optimistic may think 'this is worth looking into, I could save a lot of money here'. Whereas someone who is naturally more negative and pessimistic may think 'that's not likely to make much difference, I'd just be wasting my time if I look into that'.

The way for us to mitigate against these individual factors, and to create easier comparisons, is for us to make the benefits more tangible.

That way, they have a more accurate sense of what they could gain, and it therefore becomes an easier decision to make. Do we want to carry on as we are or is it worth us looking into this further?

In the second example, we have gone some way towards helping them. However, the reference to percentages still means they have to do some work. It relies on them being familiar with what their current absenteeism rates are and what that translates to in terms of costs to the company. If they are not familiar with these figures, they will need to find out what they are, before a comparison, and ultimately a decision, can be made. Do you think their limbic brain is going to put the effort in to find out what those numbers are? No, in most instances, neither do I.

Therefore, we need to do what we can to really quantify the change they can expect. We need to provide tangible elements that facilitate their comparisons, and support their decision-making processes. We need to show them not only what they stand to gain, but also that we have experience and evidence to prove the difference we make. The more we can quantify the results they are likely to achieve, the fewer barriers they will have to making their decision.

Now, let's be honest. I appreciate that this is not always easy for us to do. But if we don't put the effort into it, we are expecting our prospects to. And they have far less motivation to do that than we have!

So, how can you go about being able to quantify and make tangible the differences that you can make? Do you have case studies you can feature, data you can mine or testimonials you can explore? Again, this may be the sort of information you can gather when you talk to your existing customers. This is the reason I have suggested you get them to quantify the situation they were in and the difference that has been made. Through gathering information such as this, you will be able to provide more tangible examples of the outcomes your prospects can expect to experience for themselves.

2. The burden of proof

So this brings us to the final piece of the puzzle. We have attempted to understand your segments' deep fears, we have turned them around and we have been able to produce some tangible figures to help your prospects understand the likely benefits they can anticipate. Now, all we need to do is show them that they can trust our claims. We need to present the proof.

As humans, we appear to be hardwired to trust others. Yes, we have the capacity to be sceptical and suspicious, but for much of our evolutionary past, we have been surrounded by small numbers of people, most of whom we know. Not any more though! This change is becoming increasingly challenging for our brains to process, as we live in larger and larger numbers, and our worlds become more and more reliant on social media. We were never designed to encounter so many strangers and to be exposed to hundreds and thousands of people each day, many of whom we only get to see from the shoulders up and in 2D form! I believe this challenge is only going to get worse as we fully embrace the potential of AI-generated images and video content too. So, what can we do to prove that we *can* be trusted?

According to a theory put forward by neuromarketers Patrick Renvoise and Christophe Morin there are four categories of proof that we can offer by way of supporting the assertions that we make.[2] Interestingly, though, they appear to occur in a distinct hierarchy too, with some holding much more neuromarketing 'clout' than others. In reverse order, they are as follows.

Aspirational proof

This is where we talk about the way things could be. We don't have any real proof to back it up yet, but we have intentions and aspirations. Maybe we are a startup who does not have a track record yet, or maybe we are teasing the launch of a new product or service. Either way, if all we can do is talk-up the vision we have, then we can provide that as a form of proof. We need to accept that it is a weak form of proof, though, but in the absence of anything else, it might work if we bring infectious levels of conviction and enthusiasm to it.

Analytical proof

This is where we use data and statistics to prove the assertions we are making. 'Hold on a minute,' I can hear you saying, 'data and statistics are only in third position?' Yes, they are. For example, stating that '67 per cent of 182 people rated this better than their current supplier' is encouraging but hardly compelling. Why? It doesn't move us does it? It doesn't connect with us on an emotional level. It is cold and very distant. It may be better than aspirational proof, but it is not saying much in real terms. We can do better.

Observational proof

Can you demonstrate or show the difference that can be expected? This may be by using 'before and after' images where the transformation is easy for everyone to appreciate or it could be through using storytelling to demonstrate a scenario where the decision to work with your organization led to some inspiring outcome. At the moment, we place great reliance on differences that we can see and which resonate with our own situation, so what do you have that you could use here?

Social proof

This is the top, the pinnacle, the zenith of the proof tree. In social proof, we find a very compelling and reliable way for us to evidence the claims that we want to make. The reasons for this are three-fold.

First, when we are able to see that other people have made the same purchasing decision as we may do, it feeds our sense of belonging. It is almost the herd mentality in us that is reassured by doing something that is not different from what others have done. Clearly, this phenomena is closely related to the number of people who have made similar purchases and choices, so larger numbers are generally more compelling.

Secondly, if these people have positive things to say or ratings to share, then we will again gain confidence from their experience. Think about what this means for reviews like those found on Trustpilot, Amazon, Google, etc. If 2,375 people have reviewed a restaurant, and on average they give it 4.6 stars, you will gain a lot of confidence from this. Even though you have probably never met any of those people and have no real sense of what their quality standards are likely to be in relation to your own. But still, we would be very reassured by that.

The third reason this social proof holds such power over us is because it is objective. It is not an organization making claims about itself, but a third party who is making them. This again means we are more likely to trust their comments and experiences than those provided by the organization about itself. The effect of this can be compounded even further if the endorsement is provided by a known or recognized authority figure within the industry. Maybe it is a household name (individual or brand) you have supplied, an expert endorses your work or a thought-leader in your industry promotes you, etc. Each of these create even greater leverage as a result of the reputation and influence of the individual or organization concerned.

By applying these steps to your target segments and audiences, you will be able to shape and craft more meaningful messages around your offer. You will demonstrate that you know and understand their situation, their challenges and their concerns, and you will help them to see that you can be relied on to help solve them. This will all be done in a way that provides the maximum credibility and confidence too, thus building the brand and its reputation simultaneously.

> Now, there is one caveat I need to discuss with you before we move on from this section. This applies if you are in the business of selling very exclusive, premium priced products or services. Here, the 'herd mentality' approach will not serve you so well, as clearly your customers want to feel that they are in the minority, and that they are able to benefit from or experience something that most people cannot. Therefore, in this instance, the greater leverage will come from endorsements by people they look up to or even hinting that that person/organization may want it but cannot have it if it is bought by you now. In all other transactions, numbers add up. But if you work with exclusive, premium and elitist customers, do not adopt that approach.

Now we have achieved greater clarity on our key approaches and messages, let us explore some of the mechanics of how we can convey them to our audiences.

Chapter summary

1. Neuromarketing may mean we need to consider a new approach to segmentation and personas for our customer base.
2. We know we cannot rely on what people tell us about their attitudes, motivations and intended actions.
3. However, we can still glean valuable information from talking to them about their circumstances and experiences, provided we are aware of its limits.
4. Using the language of your customers will remove barriers that prevent new customers engaging with you and your materials.
5. We need to drill down to the base fears people have – security, status and strength.
6. These categories are relevant for B2B and B2C businesses, although with B2B we may need to acknowledge the additional layer for the individuals involved.
7. We need to make decisions easy, by making the benefits as tangible as possible, so the audience doesn't have to work hard to understand what they will gain.
8. When proving our claims, there is an apparent hierarchy within our brains which we need to understand, going from social, observational, analytical down to aspirational proof.

Notes

1. Kahneman, D and Tversky, A (1979) Prospect theory: An analysis of decision making under risk, *Econometrica*, 47, 263–91
2. Renvoisé, P and Morin, C (2012) *Neuromarketing (International Edition): Understanding the buy buttons in your customer's brain*, Thomas Nelson Publishing, Nashville

8

Pictures really do paint a thousand words

Context

Once we are clear on who our audience is, and what their true motivations are likely to be, we can then start to produce materials that will appeal to them. Now, I am sure when you do this usually, you begin with the words. However, for reasons that will become clear soon, I am encouraging you to do things differently from now on. I want you to start with the image in mind...

Think about a sunset. Or an elephant. Or success. Or the Taj Mahal. Or your dinner last night. Or love. For any of these, when asked to think about them, did your brain provide you with a list of words that you associate with each of them? Or did you find that you quickly had an image leap into the front of your mind? For most people, it would've been the images. Even for something as complex as 'success' or 'love', we create an image of what (or who) that means to us.

Your brain is capable of processing images at a phenomenal rate. It had previously been thought that images could be processed within as little as 100 milliseconds (10 images per second). However, research carried out in 2014 found that under some circumstances, your brain can process images in just 13 milliseconds (almost 77 images per second).[1]

Just think about that – 77 images per second. This shows us how well developed our visual systems are within our brains. Seventy-seven images. Per. Second.

Compare that with the speed at which we process words. A thorough review of research in this area concluded that for the English language, words are read at an average speed of 238 words per minute if read silently (3.97 words per second) and 183 words per minute if read aloud (3.05 words per second).[2] That is still very impressive, but you have to admit it is in a different league from the speed at which we process images.

You see, throughout our evolutionary past, we have relied on visual systems to give us information about our environment. The occurrence of words and written language as part of that is only relatively recent (more on that in Chapter 10). Images are much more natural to us, and significant proportions of our brain are given up to the practice of interpreting them.

When we 'see' something, our experience is the result of a very complicated process, involving many different areas of our brain. It is not like a camera that merely captures images that are projected onto our eyes. Oh no, far from it.

If you recall your biology lessons from school, vision is created when light waves hit the retina at the back of your eye. More specifically, the surface of your retina is made up of millions of microscopic photoreceptors. These become activated as the light waves hit them, and they convert this into an electrical signal. The signal then gets passed down the optic nerve, all the way to the occipital lobe at the back of the brain. It is here that most of our visual processing occurs. You will learn more about these photoreceptors when we move on to explore colour perception but, for now, just know that the information being presented for processing is made up from receiving millions of component parts.

These different parts are processed in different areas of the brain, and then brought back together to make sense of the overall picture. All within 13 milliseconds apparently! Isn't your brain amazing? Within these processes, there are some interesting anomalies which it is helpful for us to know about and understand. Some of these anomalies contain real potential for us in terms of how we get our messages across to our audiences.

Efficiency

The location of the photoreceptors on the retina of our eyes provides a very efficient system for supporting our understanding of the environment around us. For example, look up from the page now, and fix your gaze on something stationary in front of you. It might be on the wall, out of the window or just an item that is in the room with you. Now, as you look at that item, I want you to notice how detailed your information is regarding it. You will be confident about its colour, edges, texture probably, too. However, whilst you keep you focus on the same point, I want you to move your awareness to something in the periphery of your visual field – something you can only just see out of the corner of your eye. Remember, I don't want you to turn and look at it, I just want you to tune into its presence there.

Do you notice how reduced the information is that you are receiving about this item? The colours will be nothing like as vibrant or distinct, the edges may only be discerned as a result of changes in colour and the texture becomes very hard to determine (unless we rely on our memory to fill in the gaps for us!). Now, if you turn your head to look at this item fully, you will once again be able to see it in all its glory. Ta da!

The point of getting you to complete this activity is because I want you to appreciate that not all aspects of our visual field are processed to the same levels of depth. In the centre of our visual field is an area of the retina called the fovea. Here, you have the highest concentration of photoreceptors, and, in particular, those that are responsible for detecting colours, providing high levels of detail, etc. Conversely, as we move further away from the fovea, these aspects are no longer so strong, allowing for other strengths to come into play, such as the detection of movement etc. How much more useful is it for us to be able to tell if something is moving out of the corner of our eye than the precise details of what colour it is? This early alert system has no doubt been integral to our survival for thousands of years, and it still serves us well today.

Expectation

Much of the way our visual systems are able to operate as quickly as they do is based on the idea of our expectations. We become very efficient at processing what is in front of us because we have high levels of predictability that we can usually rely on. So, if we are entering a café, we do not need to work hard to process and make sense of most of the items around us – they are what we would expect to see in that sort of environment. Chairs, tables, cups, decorations, people, menus, etc. It is as if we are already 'looking out' for these things, we are expecting to find them, so the information from our visual system seemingly almost confirms their existence and location. This is much more efficient than starting from scratch every time we open our eyes.

Contextual information

Taking this to the next level, we need to consider how significant this role of context is within our visual processing. For example, what is this in Figure 8.1?

FIGURE 8.1 Image device for you to try to identify or interpret

OK. Assuming you have come up with an answer to that question (and I hope you have, otherwise I feel like I am here just talking to myself), what else could it be? How many different answers can you come up with for this? Three? Four? More than four?

I've listed some of the possible answers I have come up with, but please do let me know what more you have thought of too.

Can you see how hard that was when your brain had no context to provide clarity? In each of the examples I have provided here, the same device makes perfect sense. But the device has not changed at all. It is precisely the same in each presentation. Our brains interpret it differently in each case though, as the context around it gives us information and, ultimately, clarity.

FIGURE 8.2 Some of the possible ways the device featured in Figure 8.1 could be interpreted, with additional context provided

Faces

One of the ways that context is so critical to us is because our brain is essentially looking for shortcuts or those heuristics again. Identifying patterns within the visual information we receive allows us to optimize the process and keep it as efficient as possible. One of the most important patterns we are primed to notice and attend to is that of a human face.

As we have learned, within the brain one area (called the fusiform gyrus) is renowned for its role in processing information relating to faces. It is essentially always on the lookout for the configuration of elements that make up a typical human face – two eyes, a nose and a mouth – positioned in the typical way relative to each other. This has served a strong evolutionary purpose throughout our history, as it would enable us to not only identify a face/human within our environment, but go on to quickly understand if they are already known to us or not (friend or foe?) and also read these features to understand their likely emotional state too.

So powerful is this quest within the fusiform gyrus that sometimes it gets things wrong. Sometimes, we detect faces where there aren't any. We misinterpret details in our visual field, and interpret them as being a human face, when they are in fact not. We learned about this in Chapter 4, as we considered how critical our processing of emotions is. But can you see now how this comes about? The fusiform gyrus is permanently processing incoming visual information for items that present in the formation and relationship of key features of a face. And where it becomes triggered, we infer the existence of a face and even attribute a character, name or personality to the item in question. Which leads me on to...

Illusions

Some of the many flaws in the way our visual information is processed and compiled can best be seen by considering some of the optical illusions that have been created and demonstrated for many years now. Consider the example in Figure 8.3.

Such illusions work because they rely on key aspects of our visual system, aspects that 99 per cent of the time serve us very well. However, they do show that these assumptions are just that – assumptions based on experience and prediction rather than just the incoming information.

FIGURE 8.3 Optical illusion showing the effect perspective has on our judgement about sizes. The posts we perceive as being further away are judged to be taller; however, all three are actually precisely the same size

Created by Cibo00, Openclipart.org (27 August 2025)

2D vs 3D

With so many of us working online now, staring at screens for work and pleasure, it is worth exploring how the brain responds to these different formats. I am sure you can easily appreciate that 2D is a new innovation for our brains, and the speed at which we have migrated everything onto these platforms far exceeds our evolutionary adaptations. So what impact does this have on our audiences?

Research published in 2025 explored the differences between how human brains processed three variations of the same objects. One was in 2D (an image presented on a computer screen), one was virtual 3D (using a VR headset) and the last one was true 3D (in real life).[3] Among other interesting nuggets, their findings indicate that the brain struggles more to process 2D information than it does for both virtual and real 3D formats. We are used to processing in three dimensions, and to lose the insights and information that depth provides us with creates challenges and increases the cognitive load that our brains have to apply to making sense of the material. Effectively, we are making our customers suffer in order to access 2D content.

What does that mean for your online strategy?

Before you answer that, I just want to add some further insights to the case. I may be going off on a bit of a tangent here but I think you will find it interesting, so stay with me. When we spend a lot of time looking at screens or reading documents, we suffer. We actually find it very stressful, and this is more than just the reason outlined above. You see, the brain responds differently to things that are close to us, as opposed to things that are a long way away. Again, these changes are subconscious, but their impact can be life-changing.

For our ancestors, life was different. They would have spent much of their time outdoors, in open spaces, where it was possible to see things that are far away. This would have been relaxing for us – being able to see long distances, and know that there was no imminent threat approaching. On the other hand, when we were focusing our visual systems on something that is close to us, we needed to be alert. The chances are whatever is close to us demands our attention. It might be something we are preparing or making, it may be someone we are talking with or it could be something that is near to us. In any case, if we are looking at something that is within a few meters of us, we need to be vigilant and prepared to react at short notice.

As a strategy, this would have been effective for us 10,000 years ago. But, today, it could be devastating. We are not designed to maintain a state of heightened alert for more than a short time, so when we do have to, our brain and body start to suffer. We are effectively creating a state of stress, which shuts down our digestive functions, restricts our ability to think clearly, remember accurately and make good decisions. Not a state any of us would seek to be in for any length of time. And yet we are.

Without us having to do anything to trigger it, our system activates this same response just as a result of the fact we are focusing our vision on something that is close to us. Yes, that adaptive mechanism that evolved to protect us is now doing significant harm. You see, this change is triggered automatically by the change in the angle of our eyes. Humans have binocular vision – two eyes that work together to focus on the same thing – and when the object of their focus is close, the eyes are turned sharply inwards towards our nose. Alternatively, when we focus on something that is further away or on the horizon, the angle of our eyes is lessened and the heightened state is reduced or removed. So, spending hour after hour each day working on screens, having online meetings and scrolling through our devices, actually does us much more harm than we had originally thought or understood. This is also another of the reasons why the Covid-19 lockdowns were so distressing for people – our ability to view things at a distance, such as when we are driving or going out for the day, was severely compromised.

I hope you know me well enough to know that I am not a negative person. However, I am telling you all of this as it is important to realize that few of us – me, you, your family, colleagues and probably your audience too – are operating without this increased state of alert. We are therefore in a stress response before we even consider the subject of the online meeting or the content in our social media feed. So, there are two things you need to do now that you know this:

1. work hard to protect yourself from experiencing this for any extended period (and those around you who you love and who listen to you)
2. work hard to make sure you support the brain processes of your audience, as they are already challenged enough

Which brings me full circle, back to the concept of using images to their maximum effect.

We need images to do the talking for us. We need them to capture attention, tell a story, pique curiosity, create a connection or demonstrate relevance. The more of this that images can establish, the less burden there is on the audiences' cognitive processes. So, please do not spend lots of your time creating beautiful copy and writing content that flows and engages, only to then grab a picture at the last minute, while you mentally think 'that'll do'. No it will not. If the picture is not enticing in any way, the text will not get read. The picture has to convince our brain that it is worth applying the precious resource needed in order to read the words that accompany it.

Contrast

One of the best ways that we can do that is by using our images to convey a contrast. It could be a contrast between:

- us vs our competitors
- life now vs life in the future
- maintaining the status quo vs doing something to address the issue
- injustices and inequalities (for example one 'normal' vs another)
- what people are expecting vs what we deliver

Think about the classic format for washing powder adverts or stain removal liquids etc. These often show the difference between the original dirty item (complete with a devastating stain) and the new, sparkly clean one.

Or the campaigns promoting weight loss that feature the way someone used to look alongside how they look now.

Or the behaviour of a dog who rejects their current brand of food but devours the new tastier one.

Each of these tells a story. No words are needed. They use a visual representation to convey many of the benefits that can be expected. And inherent in that you can easily anticipate the way it will feel to experience those benefits for yourself. The dopamine is being released as we speak!

So, what could that look like for your products or services? How could you use contrast within your imagery to convey your most powerful messages?

Personalization

Another way that images can be very engaging and convey messages directly into our audience's brain is by adopting the technique of personalization. This works so well because once again we are making it easy for the brain of our audience to see the benefits that they can expect. However, when we use personalization, we are allowing them to literally see themselves as things could be. So, some of the best results can be achieved by using imagery which either:

1. shows the benefits from their perspective, or
2. puts them literally in the picture

Let's explore these in more detail.

SHOW THE BENEFITS FROM THEIR PERSPECTIVE

When we do this, we use images that are taken from the view of the customer. So, if we want to market mountain bikes, we don't begin by showing images of the bikes taken from the side. This is not the view anyone will have when they purchase the bike. Instead, the view to use is the one that would feature two hands on the handlebars, and the front wheel between them, as if we were heading down a forest trail. *This* is the view the purchaser will have. Showing this image makes it very easy for the busy, overwhelmed brain to be able to imagine themselves in possession of the bike, taking on new trails and living their best life.

Similarly, if we are marketing courses or a university offering, we would not want to use images that show large, usually empty, auditoriums and lecture halls. And, even worse, do not take that photo from the front of the room, because for the majority of brains that is one of their worst fears. That will create a very strong negative emotional response, which is probably not what you want them to associate with your brand. Instead, get the photographer to take images from a few rows back, sitting in a seat, with other students around them. This view, complete with laptops open and someone at the front delivering their content, is what the brain wants to experience. This is how it imagines the experience will be.

Showing images taken from these perspectives means we have done the hard work, so the customer's brain does not have to. We are not asking it to be able to make the leap from what it is being presented with to what it will actually experience. We are making that leap *for* them, and allowing them to cut straight to the part where they get excited about how that will feel.

PUT THEM LITERALLY IN THE PICTURE

The other approach I am suggesting is one I introduced earlier. It is a subtle shift, which can be very effective. It is to include the presence of the audience within the image too. So, the hands on the handlebars for instance. This is a way of literally placing the viewer into the image. At a subconscious level, our brain connects more strongly with these images as they again represent the reality we see most of the time. I am used to my hands being in the picture, as it were. At this moment, as I am typing this, I may be looking at the screen and focusing my attention on the words that are appearing on the page. However, I can also draw my attention to the location and vision of my hands on the keyboard. These were always here but, again, not usually considered to be important and noteworthy because they are often 'there'. It's like the way you don't notice your nose being in your view all the time… unless someone mentions it… which I just did… so off you go now and just check it is still there! Go on, close one eye at a time and make sure you can see it, then open them both and see that it is still there.

Have you done that? Great. So can we carry on now?

There are some brands who are using this very effectively, and have been doing so for quite some time. I am sure you can think of a particular travel company who use the imagery of a person who is effectively 'pulling' you through the holiday. They appear to be holding your hand (oh, maybe that reminds you of the music they use too!) and leading you through the airport, through the hotel, down onto the beach, by the pool, into the restaurant, etc. The interesting and clever aspect of this is that they have different versions of this device, featuring different people doing the 'leading'. In one instance it is a young child, but clearly that is not going to resonate with everyone and reflect their lived experience, so they have other versions that feature significant partners of varying ages and genders. Through doing this, and placing the adverts carefully on social media and TV channels, they can select the version that is likely to be most effective for the demographic and audience make-up at that time.

I need to issue a word of warning about this technique though. If you are going to do it, make sure you do it properly and with a very objective and critical eye. You see, the benefits of this method can be very positive and build connection and reach in a way that has not been seen before. However it can also create negative responses too. For example, using an image that contains someone holding an item, say a hot drink, in their hand could be enticing. It could stimulate responses of interest, valence, desire, etc. Or with just one artistic change, it could stimulate responses of fear, disgust and aversion. What change could create such an apparent shift? Simple – the hand that is holding the hot drink.

You see, 90 per cent of the population is right-handed.[4] Therefore, their lived experience is most likely to be holding a hot drink in their right hand. If you carefully select a beautiful image, with a delicious looking drink, and a welcoming background to make it feel accessible, but have the left hand holding the drink, it will trigger negative response in most of your audience's brains. So, please pay attention to the details, because they are likely to get noticed at a subconscious level if not a conscious one. And sometimes that can be enough to put people off. They may not be able to tell you what it is or why they don't like it, but if we had an EEG headset on them, we could show you (and them) the difference it makes.

Such is the power that images have. Now do you see why I am so keen to encourage you to place greater weight on them when you produce materials? They not only determine whether your written content gets read or not, they can also create strong associations and perceptions at a subconscious level, which all go on to inform future decisions and actions. Exciting isn't it?

Chapter summary

1. We process images much faster and easier than we process text – up to 77 images per second compared to approximately 4 words per second.
2. The images we perceive are in fact made up of millions of separate bits of information being processed simultaneously within our brain.
3. What we see is significantly influenced by context and our expectations of what we are likely to see.
4. Knowing that we have a dedicated part of our brain that is permanently looking for faces allows us to use people within our images to the greatest effect.
5. Our brains find it easier to process 3D information than the 2D versions we encounter so much in our lives today. This creates stress and significantly compromises our higher executive functions.
6. One of the easiest ways to use images to convey your message is to use it to demonstrate contrast.
7. To make images easy for the brain to anticipate themselves experiencing the benefits, we can use personalization. This means taking images from the user's perspective, and even placing parts of them into the image.

8 We need to be very careful and mindful about the images we select – they are very effective at creating unconscious responses and associations but we need to ensure we do all we can to make them positive ones.

Notes

1 Potter, M C et al (2014) Detecting meaning in RSVP at 13 ms per picture, *Atten Percept Psychophys* 76, 270–79, https://doi.org/10.3758/s13414-013-0605-z (archived at https://perma.cc/QPV7-PHGP)

2 Brysbaert, M (2019) How many words do we read per minute? A review and meta-analysis of reading rate, *Journal of Memory and Language*, 109, https://doi.org/10.1016/j.jml.2019.104047 (archived at https://perma.cc/9TFB-CBHY)

3 Kisker, J, Johnsdorf, M, Sagehorn, M, Hofmann, T, Gruber, T and Schöne, B (2025) Visual information processing of 2D, virtual 3D and real-world objects marked by theta band responses: Visuospatial processing and cognitive load as a function of modality, *European Journal of Neuroscience*, 61 (1), DOI: 10.1111/ejn.16634 (archived at https://perma.cc/77JJ-B4BU)

4 de Kovel, C G F, Carrión-Castillo, A and Francks, C (2019) A large-scale population study of early life factors influencing left-handedness, *Scientific Reports*, 9, 584, https://doi.org/10.1038/s41598-018-37423-8 (archived at https://perma.cc/J3M9-DXDC)

9

Colourful choices

Context

This is one of the areas of neuromarketing that I probably get asked the most questions about. What colour should my logo be? What colour gets people to act? What colour should I wear to my interview? It seems as though colour is a topic that many people understand has opportunities, but few people know the reality of them, or how to approach harnessing them.

Colour perception is a complex area of neuroscience. There are some aspects we know, understand and can objectively measure, and there are some aspects that we still cannot. So, please be patient if this chapter does not contain all of the answers that you wanted it to.

It starts with the eyes

You may already have come across the statement that colour does not exist in the real world. Our perception of colour is actually generated by different wavelengths of light. These bounce off the objects we are surrounded by and are detected by the retina when the light hits the surface at the back of our eyes. Remember that process, because we will come back to it later and learn how complex it can make things for us. For now though, let us go back to understanding what happens when light waves arrive at the retina. Specifically, this area is covered with millions of photoreceptors, which are classified as being either rods or cones.

Rods

Each one of our eyes has approximately 120 million rods on its retina. There tends to be a greater concentration of these around the periphery of our visual field, as they are particularly good at detecting movement and give us vision in low levels of light. In fact, they are 500 to 1,000 times more sensitive to light than their counterparts – cones.

Cones

Each human eye has approximately 6 million cones on its retina. There tends to be a greater concentration of these within the macula, as they are what give us clear and detailed visual information about our environment. It is also these that allow us to see colour. The different wavelengths of light energy are picked up by the cones, and each cone is sensitive to specific wavelengths. So, some will be activated by the long red wavelengths, other cones will be activated by the medium green ones and others still will only be activated by the short blue wavelengths. The combination of cones that are activated, along with the intensity of the activations, are what our brain then translates into the perception of colour.

All of this may be mildly interesting to you so far but what we are just learning to understand and appreciate is that when the messages are received within the brain, and our perception of colour is created, other aspects within the brain are activated too. Hence, why our perception of colour can create changes within us physiologically and emotionally.

This opens a significant opportunity for us, both personally and commercially. Just think, if you use it intentionally, you can begin to influence how you and others feel, how you are likely to act and even some of the decisions you are likely to make. Hasn't that just added a whole other level of complexity to what colour you should decorate that room?!

Evolutionary background

If we consider the process of colour perception, and how it has evolved over the generations, we can see that it has consistently served a valuable purpose. We needed to be able to determine if fruits were ripe, crops were ready, insects were dangerous, water was safe and even if meat was cooked through. We needed to be able to differentiate between plants that would heal us, allow us to create structures or provide tools to facilitate us. All of which are possible, but much, much harder if we are not able to rely on colour information.

I believe that looking at colour in this way allows us to understand some of the basic biological processes that happen when we experience certain colours. For example, when we see red, we become stimulated or aroused. This can be in terms of physiological attraction and desire (think ripe fruit and berries) but it is more usually as a result of sensing danger (think blood or hazardous animals). Either way, red gets noticed, and in noticing it, changes begin to happen within our bodies.

Whereas green has a totally different effect. If you think about it, green would have meant we were in a time of plenty, with lots of food and resources around us. Hence, when we perceive green now, it usually prompts us to feel relaxed, calm and generally at peace.

I know, I know, you want more information on all of the colours don't you? Well, far be it from me to get in the way of your curiosity! Let's take them one at a time, and see what we can discover.

Red

Associations – red is very often associated with strong emotions such as passion, love, anger, etc. This is because it gets easily noticed, excites the body and prepares it for action. It is therefore also used to depict danger, strength and dominance. Due to its ability to make people behave in impulsive ways, this colour is associated with aggression and violence too.

Physiologically – when we are exposed to the colour red, we become more alert and aroused, ready to act. Our heart rate and blood pressure increase, ensuring we are equipped to take on whatever danger, threat or opportunity approaches us. Our reactions become faster, our senses are heightened, our appetite is increased and overall we become highly stimulated. We are driven by a sense of urgency which, accompanied by increased strength, compels us to act. However, it is not all good, for in the midst of all these changes, we also reduce our ability to think analytically.

Uses – for all of these reasons, it is not surprising to note that red is often the colour most used to grab attention. It is used for road signs, sales promotions and call to action buttons.

Examples – due to the stimulation red provides and the increased appetite it creates, many fast-food companies use this as the basis for their brands and colour schemes. Also, think about the way it is dominantly used by Levi's, Virgin and Red Bull, to say nothing of the many sports teams who choose to harness its effects.

Orange

Associations – orange is typically associated with still retaining the energy of red but utilizing it in a more playful manner. Created by bringing yellowness into the reds, it uses its natural vibrancy to depict creativity, optimism, informality and fun. On the negative side though, it can be connected with feelings of arrogance, pride, overwhelm and, in some cases, superficiality.

Physiologically – in many ways, the colour orange really is made up from the sum of its parts. It continues to stimulate appetite due to the red wavelengths contained within some shades of the colour and it can be correlated to an increase in energy levels.

Uses – this is still used to capture attention but demonstrates brands who do not take themselves too seriously. It brings in elements of warmth and friendliness, so is often favoured by service providers who want to instil such feelings in their target audience.

Examples – the brands who have the strongest use of orange in their colour palettes are often ones that want to present youthful, creative and fun elements. Think Headspace, EasyJet, Etsy and B&Q.

Yellow

Associations – yellow is usually associated with very positive feelings, such as happiness, cheerfulness and warmth. It is considered a youthful colour due to its energizing and optimistic effects, and was selected for emojis as it aids the release of serotonin within our brain. However, it is worth noting that it also has negative associations such as with cowardice and deceit.

Physiologically – perceiving the colour yellow is known to increase the viewers' levels of positivity, energy, creativity and optimism. It has been shown to stimulate mental activity, concentration and analytical abilities, increase heart rate and blood pressure, and just perceiving it can also raise our body temperature. However, excessive levels of exposure to it can also create visual fatigue, and trigger both anxiety and irritability.

Uses – due to its position near the middle of the visible light spectrum, yellow gets noticed quickly. This may be why it gets used for caution and warning signs in many environments. It has also become synonymous with 'short-dated' offers in many supermarkets, creating a whole phenomenon around 'yellow sticker' offers/shelves/sections.

Examples – due to the energized and optimistic state that is created, many brands incorporate yellow into the colour schemes, although few use it in isolation as it does not always stand out against white backgrounds etc. Some of the brands who use it most powerfully are Post-it, the McDonald's Golden Arches, Mailchimp and Snapchat – all energized brands who have young and slightly disruptive attributes to their offers and brand personalities.

Green

Associations – green has very strong and clear associations with nature. More specifically, it is often linked to renewal and growth, with a large dose of harmony and balance thrown in. Think about abundance in nature, and you will get a sense of how its perception makes us feel and respond.

Physiologically – activating the green photoreceptors is known to improve mental wellbeing, creating a soothing, reassuring and relaxed state. It retains some of the yellow wavelength effects of creativity but balances them with lower levels of energy and stimulation. This results in good levels of impulse control and motivation, and can enhance our ability to focus too.

Uses – typically, green features largely when referencing healthy, natural or eco-friendly credentials. It can also be used to suggest growth and emergence – do you remember the 'green shoots' phrase that was used to excess as we emerged from Covid lockdowns?

Examples – there has been an increase in brands who have incorporated green into the palettes recently, demonstrating more emphasis on natural and environmental aspects. Think Waitrose & Partners, The Body Shop, Starbucks and Spotify, to say nothing of the icons that are now used to represent recycling and sustainability initiatives.

Blue

Associations – here we get into much calmer waters – literally. The main associations with receiving blue wavelengths are stability, peace and relaxation. We are much less likely to act on impulse, which caused an issue for Google when they learned this. You would not intentionally choose a very calm and unimpulsive colour to be used for all of your hyperlinks would you? Which is why, when they realized this, they changed the shade of their blue to one that had a more indigo feel to it, as a result of adding shades of red. This one change alone they attributed to them increasing their profits by a staggering $206 million dollars in the first year.[1] That is the power that colours can subconsciously have. And yet, many people hardly even noticed the change. Did you?

Physiologically – blue has been identified as boosting many of our functions, including productivity, communication, problem solving and innovation. Its calming nature reduces tension, anxiety and stress, and decreases our heart rate and blood pressure too.

Uses – due to its associations with calmness, communication and innovation, blue is often adopted by firms in the healthcare, tech and finance sectors. It provides a reassurance and confidence that are highly valued by brands in these areas more than most.

Examples – just think about some of the main social media and communication platforms used today – LinkedIn, Facebook (Meta), Zoom, Skype, Teams, etc. Then venture into technology – Dell, Hewlett Packard, Intel, etc. Now what about the likes of Visa, Ford and Unilever? The list of organizations who benefit from adopting and utilizing the colour blue in their dominant logos is vast.

Purple

Associations – purple is traditionally associated with royalty. That means over time it has also become synonymous with luxury and exclusivity, although elements of wisdom, mystery and spirituality can also be identified.

Physiologically – when we view colours right at this end of the visual spectrum, we find that processes such as imagination and introspection are increased. Lighter shades can provide feelings of relaxation and calmness, which often lead to lower levels of stress, reduced blood pressure and an overall sense of balance

Uses – purple is often used by brands to portray a sense of individuality, to build a unique identity and to create clear differentiation. Ironically, as a result of doing this, they often achieve a collective identity that creates community within it.

Examples – one of the most recognizable and enduring brand colour associations is with purple, can you guess what it is? Yes, it's Cadbury's. Their distinctive use of the purple colour goes back to 1914, and has been the subject of a significant trademark battle since 2004. More recent examples of brands to optimize on the inclusion of purple can be seen in Roku, Yahoo!, Purple Bricks, Claire's and Hallmark, too.

Brown

Associations – brown is, again, a colour most commonly associated with nature and organic origins. However, other associations with it include more earthy elements such as ruggedness, support and even protection. Think strong tree trunks and smell the earthy soil!

Physiologically – perceiving the colour brown provides a sense of comfort, as it is viewed as welcoming and gives us security and stability. As such, the perception of many shades of brown reduces stress levels and supports a more relaxed state. However, large quantities of brown can leave people feeling isolated and create a sense of emptiness.

Uses – adopting the more serious and reliable aspects of the colour within our brain, UK heritage signs use brown. Too much brown can feel dull and boring, therefore it is often punctuated by introducing other, fiery colours such as orange and red.

Examples – some of the best examples of the colour brown being used in virtual isolation all follow on the same theme... Nestlé, Galaxy, M&Ms! However, JP Morgan has also used it very effectively for a number of years, can you work out why?

White

Associations – white is perceived by the brain when all three types of photoreceptor cones are stimulated equally. Ironically, this creates feelings of simplicity, clarity and peace and is usually associated with purity, hygiene and cleanliness.

Physiologically – the feeling of calmness that white provides has a positive effect on our nervous systems and reduces our sense of stress and anxiety. However, it can also lead to lower levels of brain activity, resulting in reduced alertness too.

Uses – white is often adopted by medical professionals, including those in the dental and pharmaceutical industries. It is easy to see why as it provides a reassuring sense of hygiene and cleanliness, along with reducing anxiety and stress. Talk about a win-win.

Examples – typically reversed out of a block of solid colour to provide contrast, some of the best uses of white are seen in adidas, The North Face, Oreo and Apple.

Grey

Associations – grey is regarded as a neutral, balanced and safe colour. In some instances, this also incorporates aspects of authority, as the impression made is one of 'rising above'.

Physiologically – lighter shades of grey are associated with calmness and even serenity, whereas darker shades are more intense. These therefore need to be used with caution as they can create feelings of melancholy.

Uses – grey and silver colours are often selected by those industries that need to be safe and reliable, such as technology and automotive sectors. Its neutrality is often adopted by organizations, sectors and situations where objectivity and impartiality are valued. However, at its extreme, this can be interpreted as being cold and potentially also unfeeling.

Examples – Wikipedia, Rolls Royce, Sonos, Lexus and Swarovski all have primary logos that are constructed using shades of grey. Hmmm, how connected do you feel to each of these brands?

Black

Associations – black is perceived when very little light is reflected back from an object at all, therefore very few cones get stimulated and little information is received into the brain. Considered a very strong colour, black is associated with power, status and wealth. With this also come strong implications of dignity, sophistication and elegance for some products, brands and categories. However, it is also associated with negativity, sadness, death and darkness, so it needs to be used wisely.

Physiologically – observing something that is black makes it appear smaller to our eyes. This can be helpful for the fashion industry to know but is more problematic when it comes to interior design, where the perception of space is usually being maximized. On the positive side, our sensory awareness becomes heightened, but it can also prompt feelings of introspection, sadness and fear.

Uses – due to its associations with power and status, many luxury brands adopt the use of black. Additionally, it can be seen to create formality, indicate authority and even suggest an approach based on discipline or restraint.

Examples – brands who have black as the dominant part of their logo are often trying to demonstrate independence, strength and authority. They also appear to rise above trends and fads due to the timeless nature of this classic. Think about organizations and brands such as the BBC, Nike, Gucci, GOV.uk, Zara and Audi.

Pink

Hmmm, this is an interesting one. I have debated whether to even include it in this discussion of colours because in real terms the colour pink doesn't exist. It's true, I promise. There is no light frequency or length of wavelengths that correspond with our experience of the colour pink. If you picture a colour spectrum, it goes from red at the long wavelength end to blue/violet at the short end. Nowhere on this spectrum does pink appear. However, we perceive pink when red and blue/violet lights appear together in close proximity, essentially joining up the two ends of the spectrum (as happens when you see a colour wheel). So, although pink does not feature anywhere on the spectrum of light that is visible to the human eye, I will include it here... after all, you let me get away with white and black, didn't you?!

Associations – pink is usually associated with romance, nurture, compassion and femininity. It can have youthful and refreshing associations, which are deemed childish if used excessively.

Physiologically – typically, perceiving the colour pink has a calming effect on our nervous system. It has famously been used to decorate the inside of prison cells and 'away team' changing rooms, as it was believed to reduce aggression and testosterone levels.[2] However, this only works in the short-term because if people are made to stay in pink environments for too long, they become agitated instead.

Uses – particularly used for key seasons of the year (Valentine's Day, Mothering Sunday, etc) pink can lead to relaxation at the lighter shades or stimulation at the brighter shades. Too often though, it becomes a lazy device for indicating any number of female-orientated products and services. Tell me I am not the only one who gets frustrated by this...

Examples – probably the most famous shade (and use) of pink is that adopted by the Barbie franchise. However, let us also not overlook brands such as Victoria's Secret and T-Mobile who use it to good effect

Now, if only life were that simple, we would all know exactly what to do now wouldn't we? But it is not. In reality, few colours are seen in isolation, and many of our logos, environments and brand identities are made up from a number of colours. In addition to this, let us not forget that there are significant other aspects we need to be aware of here. In fact, let us turn our attention to those now.

Dependencies

The way we perceive colour can be changed significantly depending on the context in which it is viewed. Do you remember 'that dress'? You know, the one that went viral back in early 2015 because people could not agree on whether the dress was white and gold or black and blue. (Fact – it was actually black and blue.) Now, do you recall that at the start of this chapter, I told you we would come back to an important aspect of how we perceive colour? Can you remember what it was? No, don't worry. I am only using it as a small opportunity to show you how little of this content you may actually be taking in. Sighs.

When we perceive an object, we are actually seeing the light that bounces off it and using that to give us vital information about its colour. However, our brains are cleverer than that and actually put more effort into processing this information than you may appreciate. You see, there is another level of information that our brain incorporates into this perceptual experience. We know that our perception of colour is created by light bouncing off objects and our photoreceptors detecting the light wavelengths that this produces, but of course the light that 'hits' the object itself changes colour. Early in the day, just around dawn, natural light will be pinkish or yellow, it then moves to be more blue and white during the middle section of the day, before shifting again to reddish during the twilight section of a day. In all indoor environments where artificial lighting is used, be that schools, offices, shops, restaurants, gyms, hospitals, etc, the ambient lighting levels will change the way everything looks. Referred to as chromatic adaptation, our brain compensates for these changes, and uses that information to shape its processing of the colour that we end up perceiving. So, when we perceive the colour of an object, we kind of remove the effect that the lighting itself has from the wavelengths we are experiencing. This adaptation allows us to experience colour constancy in a variety of different lights and conditions.[3]

Another context variable that affects our interpretations of colour is culture. Although physiologically we generally experience colour in similar ways regardless of where we are located on the planet, what we associate with them does vary. This is as a result of the culture we are brought up in and even the language we speak.

Different cultures adopt and use colours in very different ways. For example, the colour red is associated with luck, prosperity and weddings in China, but in South Africa it is associated with mourning, sacrifice and the shedding of blood. You can see that in both cases it depicts very passionate emotions and will still command attention, but in very different ways.

If you were born in parts of Papua New Guinea, you would not have words for colours in the same way that most languages do. Instead you would refer to them as either 'dark' (to include what we may refer to as cool colours such as black, greens and blues) or 'light' (to include what we may refer to as warm colours such as reds and yellows). And if you were born into an Arctic culture, you would have many different words to refer to the colour white, incorporating variations in texture and location to allow for detailed descriptions of snow.

Individual differences

It should also be noted that our perception of colour changes according to some of our individual states. Age, for instance. As we get older, our eyes deteriorate and the lens becomes yellowed and thicker. This clearly affects our perception of colours in the blue-yellow wavelength range, but is not the only factor at play here. Our pupil size will reduce, hence we get less light coming into our eyes, and so we are again less able to efficiently distinguish between colours perceived. And finally, it appears as though our visual cortex becomes less responsive as we age, which is currently thought to be a result of our sensitivities to colour saturation reducing.

Medication is another factor. There are a large number of widely used drugs that can affect the way colour is perceived. Some treatments for epilepsy, breast cancer, heart conditions, psychotic conditions and even erectile dysfunction can all affect the way colour is perceived and processed within the brain. Although in most instances these are temporary, in some cases their effects can be irreversible, as the retina becomes compromised as a result of taking them in high doses.

Another individual element that will affect our perception of colour is our memory. There is a phenomenon called 'memory colour effect' which asserts that when we experience an object that we are familiar with we actually remember its colour, and that influences the colour we perceive it as being. We may remember that strawberries are red, so when we see an image of strawberries that is presenting us with ambiguous information, we default to stating their colour as being red. Similarly, this phenomena allows us to capture attention quickly because if we suddenly saw a red banana instead that would be strange, and our brains would be likely to be drawn to the image and want to make sense of it.

Finally, mood. Not only can perceiving colours affect and influence our moods as we have already seen, but the same works in reverse – our moods can affect our perception of colours. This is largely to do with dopamine, that neurotransmitter we looked at in Chapter 2. In addition to being implicated in our experiences of anticipation, pleasure and rewards, dopamine is also involved in the way visual information is processed, particularly colours. So, if we are feeling down and tired, and our dopamine levels are low, our experience of perceiving colours (particularly blues and yellows) will be compromised.

Of course, in reality as we go about our everyday lives, few of us are going to be exposed to just one solid colour at a time. Instead, we live in a world that is full of a rich and vibrant array of colours, all combining, complementing and competing. This variety and complexity is both an opportunity and a curse. It is up to us to influence what we can, and use all our best knowledge and insights to create intentional stages within our marketing. How can we do that? Well, the key is to be very mindful of your customer. The more we know them, understand their fears and desires, appreciate the different stages of their journey and anticipate their emotions along the way, the better we can use colour to support, reassure, encourage and motivate them. For example, using long wavelength colours such as reds and oranges works well on call-to-action buttons and to direct attention to specific areas of the page, environment, etc.

This is not undertaken from a position of manipulation, but from an intention that seeks to facilitate their decision-making processes. Overwhelming their visual senses without considering the consequences on a potential customer's body and brain should become a thing of the past. Let's never just go with what looks great, but commit to always choosing what we know will deliver results for them, and therefore us too.

Chapter summary

1. Our perception of colour comes from light wavelengths that bounce off items in our environment.
2. Perceiving different colours changes us physiologically, and therefore our moods and behaviours are influenced too.
3. The surface of our retina is made up of millions of photoreceptors, which are either rods or cones.
4. Rods are based around the periphery of our visual field; they detect movement and allow us to have a degree of vision in low light environments.
5. Cones are concentrated in the centre of our visual field, where we are focusing; they provide us with detailed images and vibrant colour information.
6. Long wavelengths of colour, e.g. reds, energize us and make us more likely to be impulsive and act.
7. Short wavelengths of colour, e.g. blues, calm us down and make us less likely to act impulsively or in a rash way.
8. Our individual experience of perceiving colour will be affected by many aspects including our mood, age and overall health, as well as differing cultural sensitivities across the world.

Notes

1. Hern, A (2014) Why Google has 200m reasons to put engineers over designers, *The Guardian*, 5 February, www.theguardian.com/technology/2014/feb/05/why-google-engineers-designers (archived at https://perma.cc/Y729-4PQW)
2. Alter, A (2014) *Drunk Tank Pink: And other unexpected forces that shape how we think, feel, and behave*, Penguin Books, New York
3. For a great explanation of this, see WIRED (2019) Why your brain confuses colors, www.wired.com/video/watch/why-your-brain-confuses-colors-color-constancy (archived at https://perma.cc/7EM7-YS4H)

10

It's only words

Context

Most of us within the marketing profession have spent significant amounts of time within our careers learning about, and focusing on, words. Copywriting is an important skill that allows us to inform and persuade our audiences, and deservedly takes up a lot of our marketing efforts. However, words are only part of what is being communicated when we hear or read some text. To understand the wider neuromarketing opportunity, we need to consider *the way* those words are presented too.

In terms of our evolutionary past, language is a very recent development. It is thought that we evolved the capacity for it between 150,000 and 200,000 years ago, so not long after the time when Homo Sapiens emerged. Since then, our use of language has significantly evolved along with us, initially being in spoken form and largely just single words. It then developed more complexity and started to become documented in symbols. The most recent addition of the written word as we know it (and as you are reading it here!) only took place some 5,000 years ago.

Unsurprisingly, throughout this time, our brain has evolved and adapted in order to accommodate (or drive?) these changes. However, today, whether we are listening to a podcast through our AirPods or reading a WhatsApp message on our smart watch, we are still using these original, language processing areas to enable us to make sense of the words. Although many of the mechanisms through which we are exposed to language have changed beyond recognition, the same core areas and processes within our brain are still enabling us to communicate and make sense of each other in this way.

However, it is worth noting that all the processes involved in the practice of reading and speech are located within the neo-cortex, that is, the new part of our brain. The older, reptilian part that drives our behaviours and decisions is actually non-verbal. Therefore, when we use words, we need to choose them very carefully, as we need them to infiltrate through to this part of our brain, and create the responses and connections that we intend for them.

So, how do we do that? Great question – I so am glad that you asked! We will come onto that later, but first, I want to share with you some interesting aspects about how the brain processes language.

How your brain processes language

Within the brain there are two areas that are key to our ability to effectively use language. These are both contained within the temporal lobe of the left hemisphere (that is, the bit slightly above and behind your left ear), and they are known as Wernicke's area and Broca's area (names after the 19th-century scientists who discovered them). Wernicke's area is predominantly associated with semantics and comprehending what has been said, whilst Broca's area is more involved with the ability to form and articulate speech.

Prior to reaching these, different areas within our brain will have been activated, depending, of course, on whether the language has come into our brains through audio channels (something we have heard) or visual channels (something we have read). We will look at hearing and audio processing within the next chapter, but for now I want to mainly focus on the way our brains process the written word – how is it you are making sense of the marks on this page?

Reading

For the human brain, reading is not a straight-forward process. Consider each of the stages required in order for you to be making sense of the words I have placed on this page. Firstly, you need the ability to identify strange squiggles against a background within your environment. Then you have to recognize familiar shapes and patterns within those squiggles which become letters. You need to understand the different sounds that different combinations of letters may make, and how those letters come together to make words. You then need to apply meaning to those words, before finally stringing those words together to make sentences and create communication.

If we need any evidence of how hard all of this is for our brains to accomplish, just spend time around someone who is learning to read. Most of us do this when our brains are young and still developing, providing us with the best resources available to enable us to develop this complex process. You see, learning to read actually changes the way our brain is structured. Research carried out by Stanislas Dehaene et al showed that when we learn to read we are essentially re-wiring our brain, and 'recycling' areas that were previously used for other things.[1] Essentially, the process of reading utilizes existing mechanisms and functionality that occur within the brain, predominantly vision and speech recognition. However, our brain also has to recycle another element to create the connection or bridge between the two. The discovery of this fact has gone on to explain many of the issues found in children as they learn to read and write, and, as a corollary, has been used to introduce change within education practices too.

As a competent reader, when you 'see' a word, the first area of your brain to become active is the occipital lobe at the back. This is the area where visual processing takes place, so it makes sense that the activity begins here. It is then passed forwards within the left hemisphere, into an area dedicated to the recognition of written words, and, more specifically, to your recognition and understanding of letters. This may feel like a simple aspect to you now, but it takes time to learn that 'E' and 'e' are actually the same letter! These letters are then given sounds, which combine to create words. Finally, the activity shifts to your Wernicke and Broca's areas, where you allocate the words' meaning, and can appreciate the way they are pronounced and articulated. Phew, that is hard work isn't it?

Over time, this process gets much easier and faster for us, and you become able to read quite complex sentences and passages in short amounts of time. This is due to our brain using saccades (rapid movements of the eye between fixation points). In order to make sense of a visual environment, the eye does not move smoothly around it as we may think. Try it now and see if you can do it... I am serious, just look up from the page and try to sweep your eyes in a continuous smooth motion across the visual field in front of you.

Could you? No, I didn't think so. What your eye actually does is create a series of jumps, moving it from one focal point to the next. If you are clever, you may have managed to move your head in a continuous, smooth way, but your eyes would still have made a series of jumps. Try sweeping your eyes smoothly around your environment again, and see if you can notice the jumps or saccades that your eye makes.

Could you see how it works now? These saccades allow you to focus on parts of your visual environment and gather critical information about it whilst your eyes are still. Then, your eyes saccade or jump to another part to focus on and the process gets repeated. It is worth noting that saccades can happen in any direction, including looking up and down, and they can take place a number of times within even one second. Impressive aren't you?

Your brain may use saccades on a large scale like this for taking in its visual environment, but it will also use smaller ones when reading a passage of text. As a new or developing reader, our saccades will be short and we may fixate on each letter contained within a word. However, as our proficiency improves, we can increase the gap between our saccades, so our reading speed increases too. And this leads me on to the next important point about our brains when we read.

Predictive text

In many ways, our brains are now understood to optimize their efficiency as a result of making informed predictions. If I open the door to my office, I do not need to apply all my visual senses to what will be on the other side of it, in near Wonka-initiated awe. No. Based on previous experience, gained from many, many occasions of opening my office door, my brain predicts what is going to be on the other side of it. And so that is largely what I see. However, as we have already learned, my brain will draw attention to things that are new or unusual in that environment – the neuronal equivalent of exception reporting if you like. So, just as your phone and keyboard now predict what they think you might be about to say next, so your brain predicts what it thinks it is going to read next. If you are not convinced. Try reading these well-known quotes and phrases:

'If at first you don't succeed, try, try, try agony'

'Curiosity killed the caterpillar'

'It cost an arm and a legume'

For each of these, hopefully your brain would have predicted what it was confident the words were going to be by the end of the sentence. However, when your saccade landed on the word, it did not fit the prediction, so the brain had to put more effort in to making sense of it. And here, my fellow marketers, lies our first opportunity.

We have created surprise. We have snapped the brain out of its semi-automatic state, and forced it to apply more cognitive effort to reading what we have written. This one small step means that we, straight away, significantly increase our chances of being recalled. Remember how attention and effort often lead the way for memory processes to follow?

Think about some of the great campaigns and slogans who have used this technique:

- 'Freshly clicked' was created by Tesco and written on the side of their home delivery vans, using the play on words to reassure customers that the quality they get delivered, would still be what they expected.[2]
- 'Shave time. Shave money' was created by the Dollar Shave Club to promote its subscription model, emphasizing not only the products they offer but the factors that differentiate them too.[3]
- 'It butter be Warburtons' featured as part of their Mad About the Bread campaign in 2023, which included a number of creative uses of wordplay such as this.[4]

We know that surprise can be effective for us in snatching attention and cognitive effort for the benefit of our content. But what other devices do we have available for us to deploy in order to be effective neuromarketers? Well, I feel that the best way we can approach this is to mirror the processing stages that the brain goes through, so I have broken the opportunities down into four categories:

1 The words we use
2 Pulling them together to create sentences
3 Pulling sentences together to create stories
4 The way words are laid out

The words we use

How much effect do you think the words you use have upon your brain? Any idea? Well, let me tell you… they literally change it. In her fabulous TED talk, Lera Boroditsky provides a number of examples of how the language we speak changes the way that we think.[5] For example, if we are raised in a language that has different words to differentiate what we refer to as light blue, from dark blue, then we actually experience the perception of 'blue' differently.

We know that individual words can change other processes within our brain too, including those that regulate our physiological and emotional responses.[6] But they can also change our behaviours and even our intentions. Let me share some examples with you to help you appreciate the significance of just how powerful they are.

Take the word 'free'. In his book *Predictably Irrational*, Dan Ariely describes some research he and his colleagues carried out to explore the impact of the word 'free'.[7] They started out by giving participants a choice between two sweets: they could either have a Lindt truffle for 26 cents or a Hershey's Kiss for 1 cent. When this was the offer, the participants were evenly split in choosing each one. However, when they reduced the price of each item by 1 cent (so the price difference between them remained constant, now at 25 cents for the truffle but the Kiss was free), they found that 90 per cent of people then chose a free Hershey's Kiss.

So, our economic decision-making may be easily influenced by the appearance of the word 'free'. But this is a small and inconsequential example, I hear you saying, surely the word 'free' is not enough to change us in more permanent ways. Is it? Oh yes, indeed it is.

In a further piece of research, Dan and his colleagues had the opportunity to monitor what happens when people are offered the chance to have a free tattoo. Now, let's just be clear, this is a permanent piece of artwork that will be applied to your skin. Permanent. So it is shocking to note then that 68 per cent of the people who registered in the queue to take advantage of this offer said they were only getting a tattoo because it was free.[8] Sixty-eight per cent. Am I the only one who finds that a worrying statistic?

What other words affect us in such compelling ways? 'New' definitely attracts our attention. This could just be as a result of our processes associating the word with novelty and something different. We know our brain naturally needs to understand and assess new things in its environment to determine if it is a risk or not. But I think it is more than that. I think many of us are subconsciously (if not also consciously!) looking for the next best thing – trying to find products and services that will make our days easier, better, more organized, more efficient, save us time, save us money, make us look younger, look slimmer, look wealthier, make us healthier, help us to be a better partner, be a better parent, make our lives more exciting, colourful or just a little more tasty! Hence we are drawn to the word 'new'. Maybe this time it really will provide the solution we have been searching for… !

Without a doubt, one of the most powerful words we can use though is simply the word 'you'. Remember how that reptilian part of our brain is inherently self-interested? Well, if we use the word 'you' in our writing, we not only speak directly to it, but we also save it from having to apply cognitive effort to understand our message. You see if we phrase a statement or claim around ourselves, us, then the audience's reptilian brain has to decide to put the effort in to turn it around so it can see what it means for them. Take the following example.

> Due to our efficient processes and logistical operations, we guarantee next-day delivery to anywhere in the UK.

Now, I can understand what that means, but it does not feel very relevant or exciting to me. Does it you? So what about if we change it round and put the customer at the centre of this? If we make it about them, using the word 'you' in a prominent way? And tell them what it means for them...

> You could be wearing your new outfit tomorrow night.

Do you see the difference? Maybe you even feel the difference within your body somehow. This will stimulate you in a way that the previous passage just missed. So, try to make your writing more about 'them' (your customers, prospects, stakeholders, etc) than it is about yourself, your brand and your organization. Turn it all round so that the impact on them is easy to identify and appreciate.

Clearly not all words affect us positively though. Hearing the word 'no' can increase activity in part of your brain called the amygdala, and, in turn, this encourages the release of stress-related chemicals within the brain. No big deal, you might think. But you would be wrong. You see, this will inhibit the normal functions in your brain that relate to processes such as reasoning, logic, language and communication. Continued exposure to negative words can affect your feelings, emotions, memories, sleep, appetite and ultimately your longevity. Meanwhile, hearing terms such as 'amazing', 'fabulous', 'incredible', 'phenomenal' or 'splendid' can lead us to doubt and mistrust the speaker or author.[9]

Creating sentences

We now know that we need to be very intentional about the words that we use. However, the way we connect them together to form sentences can also create changes within the brain and produce significantly different effects.

Do you buy that? Do you? Because you should. Honest. It's this simple. Do you feel it? Do you feel different? Now my sentences are short. Do they feel intense? What is that doing to you?

Sorry about that. But hopefully you can see that even by simply adjusting the average length of sentences we can create a shift in the energy of the reader. We can increase the chances of them being motivated to act because the text has become faster and more compelling.

ADJECTIVES

Or we can use the sentences we construct to engage their senses and get them to almost experience the product/service we are promoting. Probably one of the most effective uses of this was made by M&S when they launched their 'This is not just food...' campaign back in 2004. The adverts centred around a TV campaign that showed off some of their premium products, such as a chocolate pudding. However, the voiceover that accompanied the intimate videography of the hot chocolate pudding being cut into with a fork said something along the lines of:

> This is not just any pudding. This is a sumptuous, rich, chocolate pudding, filled with warm chocolate sauce, and served with lashings of thick cream.

The campaign was a hit, and the chocolate pudding alone experienced a 3,000 per cent increase in sales.[10] Soon, it was being parodied and imitated across many different genres, demonstrating that it really had crossed over into the stuff of legend. So, why was this? In a word, it was adjectives.

The use of adjectives that M&S applied to describe their products was tantalizing. It triggered the creation of a whole movement affectionately referred to as 'food porn'.[11] The audience could not only see the products looking appetizing but they could also hear the evocative descriptions that were being made about them. Strings of terms like 'farm assured, naturally fed, extra-succulent' to describe their roast chicken and 'pan-shaken, ready-to-roast, extra crispy' to describe their roast potatoes stimulated the appetites and minds of viewers and readers alike. M&S had used language to reach into the non-verbal reptilian brain and create desire. Things have never been the same since!

QUESTIONS

Other ways that sentences can be constructed to leverage more impact is to turn them into questions. Consider the differences between these two passages:

> Our software is proven to significantly reduce staff absenteeism, as it uses AI technology to monitor and predict optimum productivity patterns.

> Do you know what staff absence costs your organization? How would you like to save yourself that cost several times over, through investing in the latest AI technologies?

By turning the proposition into a question, the brain naturally starts to consider it and formulate the beginnings of an answer. However, there is an important fact to note about using this technique – you need to allow the reader's brain the time to answer. If you effectively 'talk over' their thinking time, you completely lose the value that you have just created. If you are delivering a presentation or creating audio or video content, remember to pause to allow their brains time to consider the question and start to answer it. If you are using writing, create some white space around the question to let their brains answer again before they move on. In any format, this means the brain is actively engaging with your content again, so pulling words together to form questions is another easy method that we can use to create connections with our audience.

An extension of this concept is using questions that actually provide the brain with a bit of a challenge. What do I mean by that? Simple. Ask a question that is in the form of a quiz or a puzzle of some sort. Start your video, copy, presentation or pitch by asking 'which is the odd one out?' or 'what do these four items all have in common?'. Doing this will make their brain curious, they will want to know the answer, and they will be prepared to put more attention and effort in as a way of finding out the answer (and of course to determine if they were right or not!). Clearly, we do not want to make it too challenging, as we risk the reptilian brain deciding it is too hard, and just passing our content by. However, the right amount and the right options will pique their curiosity, test their knowledge and give them an irresistible opportunity to prove their intellect. This in turn will make them invest, make them engage and make them hold out to get the answer. Again, you need to allow them the time and space to consider the puzzle and come up with their solution, but the trade-off for doing that is potentially vast. Particularly if you tie the puzzle to a key element of your brand, your service, etc. It will not only become very memorable, but it could even be shared among colleagues, creating your very own version of a viral trend!

THE RULE OF THREE

The final element I want to introduce you to here is the value of using the 'rule of three'. This is a well-known approach that encourages writers, presenters and negotiators to always adopt the introduction of three elements within their content. Using three examples, or outlining three options, allows the brain to make comparisons, observe trends and enhance their likelihood of remembering them all too. Consider phrases like 'Eat, Sleep, Repeat', or 'Stop, Look, Listen' or 'Hands, Face, Space'. Each of these concepts relies on the notion of three elements to make it clear, compelling and memorable. But why is that?

Three is the smallest number we need in order to be able to find a pattern. If we only have two aspects, they could be connected in any number of different ways. But add a third and the odds improve drastically. Three allows us to easily remember each of the elements, and if we can make them rhyme or use alliteration, then the chances improve significantly again. The convenience and impact of this simple device has been used in a wide variety of formats, from children's stories, pop bands, movie titles and slogans to even the British Transport Police adopting its use in their campaign 'See it. Say it. Sorted'.[12]

The other advantage that three offers us is the element of surprise. If we start a joke with the immortal line 'three people walk into a bar…', we know that we are having a situation set up where the third element will deliver a surprise – hopefully a humorous one at that! Through being able to notice a pattern, our brains start to make predictions. And we have learned already about the power of surprise when it comes to capturing attention and facilitating memorability. So, say it loud, say it proud and say it three times. Ready? Steady? Go!

Creating stories

Once upon a time there was a part of our brain that loved stories. We still have it. The end!

Throughout our evolutionary past, stories and storytelling have shaped and developed us. They were traditionally used to share information, experiences and insights across territories and generations. Travellers who came from outside of the group would share tales of the places they had visited with eager audiences who had never left the area. Elders would use stories to convey powerful messages of previous experiences and lessons hard learned either by themselves or by previous generations. Parents would use them to caution children about potential dangers and to help them differentiate between right and wrong. Religious leaders used them to build communities, loyalty and followings, spreading the word far beyond their borders and lives. The list goes on. Today, we have all of these and more – films and TV series, books, even social media platforms that specifically allow us to post our 'stories' to share them with our own modern-day followers and communities.

It is hardly any wonder then that stories activate our brains in compelling and exciting ways. For a method to have endured and survived so much of our evolutionary development, it must offer significant advantages and benefits. However, now we are able to view it through a neuroscience lens, it seems as though these different forms of escape and education that we have relied on for generations deliver more than we could ever have imagined.

First, if the stories we are told contain emotions, they are more likely to encourage us to act. Presenting someone with data on a situation is unlikely to bring about significant change. However, stories can, and frequently do, do just that. If you tell me a story, I can relate to the main characters and imagine their feelings. I can put myself in their shoes, and that is much more likely to affect me in a way that is strong enough to create change.

When you experience a story, your brain actually does just that – it experiences it. Literally. You may find yourself sitting on the edge of your seat or feeling the same sense of fear or relief as the protagonist does. This is as a result of a phenomena called neural coupling. If the story is told well, your brain will essentially mirror that of the storyteller. This not only means you feel the richness of the tale unfolding in a more exaggerated way, but you are also more likely to recall it too. Essentially, experiencing it in this way mirrors the encoding of explicit episodic memories such as we learned about earlier. That way, we can more easily recall and impart the tale again to others within our spheres.

In addition to how we experience the story, we are likely to also feel differently towards the storyteller. This is because as we listen to them tell the story and we experience neural coupling, our brains also release oxytocin. Oxytocin, as we learned in Chapter 2, is the bonding hormone that allows us to establish connections, build relationships and create trust.

So, if you want content that is likely to actually create change, be memorable, encourage the audience to connect with you, ensure they feel something and view you with increased trust and compassion, stories are the way forward. You may have used them before because everyone is 'banging on' about storytelling, but now you can use them properly, because you understand the reasons they are such a powerful linguistic device.

Layout

I am sure you have seen content like that shown in Figure 10.1 before on social media. If not, you have definitely already seen a similar one in Chapter 3.

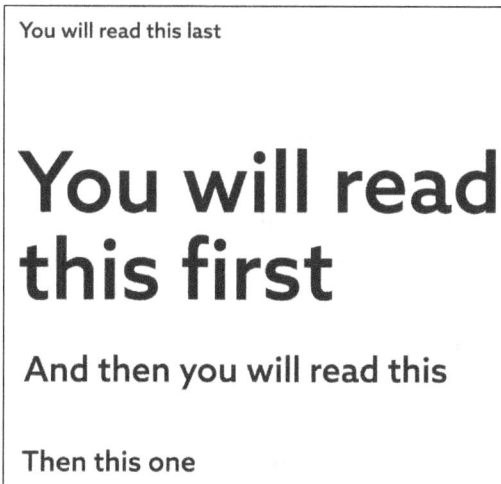

FIGURE 10.1 Example of a graphic device that demonstrates the way our eyes will navigate around a design

You will read this last
You will read this first
And then you will read this
Then this one

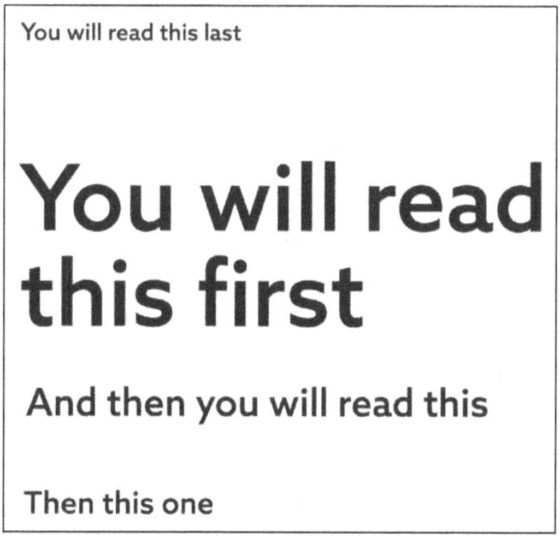

Although it is a fun piece of content to read and share, it is also making a very important point about the extent to which we can predict the way people will engage with your content. Through using eye-tracking glasses, researchers have been able to identify the extent to which content catches and then holds audiences' attention. However, we can take this a little deeper.

How much thought do you usually give to the fonts that you use? Yes, many of us have corporate fonts that we have selected to feature in the promotion of our brand and organization. But do you realize that the font itself can be a factor in whether people recall you or not?

Consider the examples of well-known (and well-used) fonts in Figure 10.2.

FIGURE 10.2 A range of commonly available fonts for you to consider and choose between

This one is Arial - would you choose to use it?

This one is Skia - would you choose to use it?

This one is Alice - would you choose to use it?

THIS ONE IS AFRAH - WOULD YOU CHOOSE TO USE IT?

This one is Arctic - would you choose to use it?

THIS ONE IS BARON - WOULD YOU CHOOSE TO USE IT?

There you have it. Six fonts to choose from. Now, which do you feel is going to be the most effective for you to use?

Have you chosen one? Which one did you go for… and, critically, why did you go for that one?

Answers to this question typically revolve around whether it is easy to read or not, with the ones deemed to be the easiest being the ones most selected. However, let me turn things on their head for you. You see, when we look at activity in the brain, we can see that, actually, choosing a font that is not so easy to read can be a desirable decision to make. If the brain has to work a little bit harder, and it has to apply a little more cognitive effort, then there is an increased chance that the content written in that font is going to be remembered. Now, please don't get me wrong, I am not saying that you want to have blocks of text that are very hard to read. No, I am not saying that at all. What I am saying, though, is that sometimes we may choose to put specific sections of text into a different font to support the readers' ability to recall it. Maybe details such as our website address or the key dates when an offer is being applied: these will be recalled better if they are in a slightly more complex font to process.

As a final note, I also want you to get good at removing words from their current context. Our brain likes to make sense of things in context but what happens if we remove that context and just look at the words in isolation? One of the best-known marketing slogans was created by doing just this. It is based upon the last words of death row prisoner Gary Gilmore just before his execution in 1977. His words, removed from their context, were adapted and presented as a possible tagline by the agency to their client. The audience loved it and it has stayed as their tagline since its creation in 1987.

Any idea which tagline I am referring to? Which three words grew from these inauspicious beginnings?

Gary Gilmore's last words were 'Let's do it'. Which Dan Wieden turned into the iconic Nike slogan.[13] The rest, as they say, is history.

Chapter summary

1. In marketing, we need to prioritize images over words as we process them so much faster – and our reptilian brain is non-verbal.
2. The ability to read the written word is very recent within our evolutionary history, and involves a number of regions within our brain being re-purposed to allow us to make sense of letters on a page.
3. When we read words, our eyes do not move in one continuous, smooth movement. Instead, they make a series of jumps, called saccades, from one focal point to the next.
4. In much of what we read, our brains predict what it expects the text to say, and that way it operates more efficiently and faster than having to read every individual word properly.

5. Individual words can change our brain and our behaviours. Using words like 'free', 'new' and 'you' all have a significant impact on the response in our audience.
6. Being intentional about the structure and design of our sentences allows us to influence the effect they will have on the reader or listener.
7. Using stories in our writing allows us to create content that is more connecting, engaging and memorable than would otherwise be experienced.
8. The font you use and the way you lay out your words will also have a significant impact on the audiences' unconscious responses and conscious behaviours.

Notes

1. Dehaene, S (2013) Inside the letterbox: How literacy transforms the human brain, Cerebrum, 7 June, https://pmc.ncbi.nlm.nih.gov/articles/PMC3704307/ (archived at https://perma.cc/XT32-PVFJ)
2. Jefferson, M (2021) 'We need to solve problems for customers': Tesco's CEO on the future of rapid delivery, MarketingWeek, 18 June, www.marketingweek.com/tesco-ceo-rapid-delivery/ (archived at https://perma.cc/FB6Q-VW4U)
3. The Marketing Society (2016) Digital drives a richer customer experience for Dollar Shave club, 30 October, www.marketingsociety.com/the-gym/digital-drives-richer-customer-experience-dollar-shave-club (archived at https://perma.cc/4V3B-DSGY)
4. Noble, J (2024) Warburtons mad about the bread, British Brands Group, 15 April, www.britishbrandsgroup.org.uk/2024/04/15/warburtons-mad-about-the-bread/ (archived at https://perma.cc/UTL3-B379)
5. Boroditcky, L (2018) TED: How language shapes the way we think, www.youtube.com/watch?v=RKK7wGAYP6k (archived at https://perma.cc/K9WS-XRYD)
6. Newberg, A and Waldman, M R (2012) *Words Can Change Your Brain*, Hudson Street Press, New York
7. Ariely, D (2009) *Predictably Irrational*, HarperCollins, New York
8. Ariely, D (2010) The power of free tattoos, Dan Ariely Blog, 10 November, https://danariely.com/the-power-of-free-tattoos/ (archived at https://perma.cc/NC69-F4WF)

9 Newberg, A and Waldman, M R (2012) *Words Can Change Your Brain*, Hudson Street Press, New York

10 Marks & Spencer (n.d.) Timelines, M&S Archive, https://archive.marksandspencer.com/timeline/advertising-branding/ (archived at https://perma.cc/K473-VDY8)

11 Barker, M (2020) 'I want people to lick the telly': Inside the launch of M&S's 'This is not just food, MarketingWeek, 18 March, www.marketingweek.com/inside-story-marks-and-spencer-food/ (archived at https://perma.cc/FVF4-KJ8T)

12 British Transport Police (n.d.) 'See it. Say it. Sorted', www.btp.police.uk/police-forces/british-transport-police/areas/campaigns/see-it-say-it-sorted/ (archived at https://perma.cc/PU79-L8W8)

13 Weiden, D (n.d.) Nike (1987) – Just Do It, Creative Review, www.creativereview.co.uk/just-do-it-slogan/#:~:text=Wieden%20drew%20on%20a%20surprising,the%20impact%20it%20would%20have (archived at https://perma.cc/9MEW-6LSE)

11

Sounds good to me

Context

Most of us are in a constant state of exposure to sound. Just think about it. When was the last time you were somewhere that was completely silent? Can you even remember? If so, the whole reason you can recall it is probably because the silence was so very remarkable. Even as I write this, I am aware of the whirring of the dishwasher in the background, and a dog snoring at my feet. Silent? Far from it! So what does this mean for our marketing efforts, and what do we need to understand about sound in order to enhance customer engagement?

Ears are strange things aren't they? I mean, they are sometimes ridiculed, often pierced and occasionally referred to in Old Wives Tales as being either made of cloth or going red when someone is talking about us! When really their sole purpose is to capture the soundwaves around us, and direct them deeper into the ear where they can be processed.

Just as our eyes convert wavelengths of light into a representation of the environment around us, so do our ears. Only they are not activated by light waves, but by wavelengths of sound. These wavelengths enter our ear canal, and carry down it until they reach the ear drum. Here, their wave form causes the ear drum to vibrate, which then becomes amplified by three tiny bones within our middle ear. These don't just amplify the vibrations though, they also transmit them to the inner ear and in particular to the cochlea, where they are converted into electrical signals. Similar to visual information, these are once again communicated to our brain for processing, but this time via the auditory nerve pathway.

Unlike the location of the processing of visual information within our brain (predominantly in the occipital lobe at the very back of our brain, far away from our eyes at the front), our auditory information is processed very close to the ears themselves. On each side of our brain, we have an area located just behind the ear, where most of the core processing takes place. Convenient, eh? So, what happens there, and how can we marketers benefit from understanding the processes at work here?

For much of the animal kingdom, the ability to accurately detect, identify and respond to a variety of sounds is integral to survival. And in many ways, we humans are no different. For thousands of years, we have used sound to isolate predatory animal calls from those that are an easy food source. We have been able to communicate efficiently, identify specific people, convey emotions and even determine the ripeness of crops, all just by using sounds.

More than that though, the way we process sounds means we can also locate the source of any of these items within our surroundings. Just think about how quickly you turn around when you hear a strange noise behind you. It is instant, immediate, instinctive. But you also know, with an alarming degree of accuracy, where you are going to need to look when your head finishes revolving enough to bring the area into your visual focus. It is a skill you learn from a very young age, and you hone it by playing games like 'Grandmother's footsteps' when you were a child! This is a key feature of us having two ears, one located on either side of your head. The source of sounds can be triangulated between the two of them, which allows us to determine with a high degree of accuracy the position it came from. So, other than locate the origin of a noise within our environment, what else do our ears do when they process the myriad sounds that we encounter and hear?

Clearly, a lot of the sounds we hear are based around language. Whether it is a podcast we are listening to, a conversation we are having or even just being aware of people talking as we walk down a street, language makes up a significant proportion of the sounds we hear. The precise proportion varies significantly depending on our age, environment and occupation, but it is still going to be highly significant. I feel we have covered language and words already though, so for the rest of this chapter, I will be concentrating on the non-language sounds we hear and process.

Outside of language, there are three main categories of sounds:

1 non-verbal human sounds, e.g. laughing, coughing
2 environmental sounds, e.g. machinery, traffic, animals
3 music, e.g. well, music really!

Let us consider each of these in turn for a moment, and see what we can determine about their role and influences.

Non-verbal human sounds

When you reflect on the scope and nature of the sounds we humans make that are not speech, we can break them down into three main categories.

Physiological sounds

These are the sounds involved in our physiological processes, such as breathing, coughing, yawning, etc. Although they may have important physiological functions, they do not generally carry much significance to those around us – hence why we tend to try and minimize their occurrence and impact. However, if the coughing were to become more of a choking sound, clearly the nature of that change would indicate a need for assistance in those nearby. Can you see how even physiological sounds can convey important information?

Communication sounds

These are sounds that are designed to communicate a significant volume of information, even though they use no words. For example, when someone sighs audibly, it can be used to convey boredom or frustration (just ask anyone who shares their home with a teenager!). Similarly, groaning can convey discomfort or even pain, whilst a sharp 'tut' can communicate many levels of judgement and even dissatisfaction.

Emotional sounds

These are clear indicators of the emotions someone may be experiencing at any given time. As with facial expressions, they are usually used to inform people around you of the state you are currently experiencing. Laughing tells people that you are relaxed and at ease, whilst crying is a more direct call for help and intervention.

As humans, many of these sounds that we use and make transcend conventional language barriers. So even though we may not speak the same language, we can often still convey significant information to another person through utilizing some of these non-speech sounds.

Environmental sounds

The background noises to our lives are often made up from a variety of man-made and natural sounds that occur around us. These may be the whir of the air conditioning in the office or the laptop we are working on, the sounds of the engine as we drive or our footsteps on the floor. They could be the birds singing or dogs barking around us, the sound of the washing machine or even the air fryer 'ping' to tell us that dinner is ready.

Each of these sounds provides information to us, but many of them will be ignored, overlooked and even dismissed as being just 'background' noise. Hence the need for us to do something different and significant if we want to get attention. Consider the sound of an alarm going off – whether it is a car, burglar or fire alarm – they all have to be very shrill and enduring in order to prevent us from just dismissing them.

Music

This is a very interesting set of sounds. In many ways, music is in a league of its own, as listening to it can have highly significant effects on your brain and, as a corollary, on your physiological state. These effects include:

- **Activation:** When participants listen to music in an fMRI, we can see that many areas of their brain become activated, not just those involved with listening. Yes, the auditory areas are engaged and working but so is the limbic system (which is implicated in our emotions) and key areas involved in memory such as the hippocampus.
- **Emotions:** Music can have a direct effect on our emotions. Think about the theme tune to *Titanic*, *Jaws* or *Rocky*. Each of these pieces of music will stimulate the amygdala in different ways, creating a sense of sadness, fear or motivation within just a few bars. Maybe you have your own playlist that you like to use when you go for a run or one for when you are in need of motivation or energizing. If so, you are already harnessing this aspect of music within your daily routines.
- **Memory:** Music appears to have some very strong connections to our memory processes, which endure over a lifetime. If you think back, can you recall theme tunes or songs from your youth? Additionally, patients who live with Alzheimer's disease can often respond very positively to music from earlier in their lives, demonstrating a connection and recall which is not seen through any other experience.[1]
- **Attention:** Music has been shown to help develop our attention span and support focus when it comes to the completion of complex tasks such as problem solving and staying alert.
- **Stress:** Listening to music can lower cortisol levels within the body and therefore reduce both stress and anxiety, and increase feelings of relaxation and contentment.
- **Sleep:** As a result of the more relaxed state, music can enhance sleep both in terms of its duration and quality, providing a range of beneficial effects.

- **Pain:** Music can increase people's threshold to pain, meaning they do not feel it so intensely. This has been used very effectively to manage people with both ongoing pain conditions, and also to allow for short-term interventions without the need for such strong medications.
- **Heart:** Music can directly affect our cardiovascular health, as it lowers both blood pressure and heart rate.

Selecting the 'right' music

Now, I don't know how you feel as you read through all of this list, but clearly there is music, and there is *music*. In order to get the most of these benefits, we need to select the most appropriate music for the individual and the benefit we are trying to create. Relaxing spa music may reduce stress and anxiety in most people, whereas some intense rock or drum and bass may have the very opposite effect. So, pick your music carefully.

Which is precisely what many restaurants do. Research has demonstrated that the choice of music that gets played in a restaurant has a significant effect on the behaviour of the patrons.[2] Playing music that has a slower tempo led to diners taking longer over their eating experience and leaving smaller tips for the staff. Conversely, where the tempo was faster, the diners ate more quickly and left larger tips too. Overall, there were no differences in the size of the bill between these two groups. So, do you want your diners to sit around, enjoy the ambience and leave you smaller tips or would you prefer them to eat faster and leave larger tips, therefore potentially freeing the table up for more guests? The choice really may be yours to create.

It is also worth considering the role that music plays in terms of setting the ambience and atmosphere in any given moment. Movies and TV shows use music to great effect by adopting it to create tension, accompany action, build suspense or convey romance, for example. These play a vital role in accompanying our experience of the content, and helping us to physically connect with the situation and characters being featured. When you are next watching a film or episode of your current favourite TV show, spend a few moments watching it with subtitles and the sound off. Take time to really notice how different your viewing experience is. It is not just about the lack of voices and atmospheric sound effects, it is also the vital cues you will miss, without having access to the intentionally composed and created soundtrack.

New ways of experiencing sound

This seems like a good point to draw attention to the change in the way many of us now experience sounds. For most of our evolutionary past, sounds were communal. Everyone within the vicinity would hear the thunder or the cries of a hungry baby. Over time, our experiences have become more and more personalized. We built walls to keep our noises in and other people's noises out. We built rooms to divide up our own noises. And now within the same room, we can have several individuals wearing headphones and earbuds, all listening to their own choice of content. Or even hundreds of individuals if we are attending a silent disco!

Similarly, when people walk down the street now, fewer and fewer of them will be listening to the sounds around them. And if they are jogging or running, then almost everyone will be listening to something they feel is appropriate. Increasingly, we want to control the sounds we hear in all of the moments that make up our day.

Quite apart from the safety issues created by everyone being so oblivious to their surroundings, this change is something we need to factor into our marketing approaches too. Will this create an issue for you or an opportunity? Is the constant exposure to different forms of noise, all being consumed individually and at different times, something you can use to your advantage? Surely all those channels and platforms, all those hours of content being consumed on a daily basis, give us additional opportunities to get inside the brains of our audiences?

Well, yes, and no. Yes, we have more channels through which we can be more targeted and approach very niche segments in ways that were not available to us before. And, no, because most of this consumption now happens while people are doing something else. We are unlikely to have their undivided attention. This is not critical, as we know that information can still be received and processed below our conscious threshold. However, if we are looking to convert them, get them to complete a call to action and just have people engage with our content, this lack of focus on it can be highly detrimental to our results. So, we need to create ways to become noticed then, don't we?

Before we move on to look at how we can harness insights about our processing of sounds for our marketing advantage, what else do we need to understand? Well, remember the Reticular Activating System, the RAS as we affectionately refer to it? This is the part of our brain that decides what is important to us, and selects those aspects for us to be consciously aware of. This not only selects based on our visual information, it also manages the audio inputs we receive, along with those from all of our other senses. Therefore, the RAS will again be in control of what we hear, separated out from all of the other noises around us.

Do you also recall learning about the phenomena where you hear your name mentioned in a neighbouring conversation? Yes, that was your RAS too, wasn't it? But it can apply the same process to any number of audio elements, depending on what it understands may be important to us. Consider a new parent and their reaction time to hearing their new bundle of joy crying during the night. Or an aviation enthusiast who believes they can detect the sound of a Rolls-Royce Merlin engine, and dashes outside to look up to the skies. Or a teacher who can detect the tell-tale sound of a phone vibrating within a student's pocket, when no such devices are allowed in the classroom.

Just as the RAS decides what makes the grade and gets our attention, it also decides what we do not notice. The sounds that get lost in the background and barely even register with our conscious processes. It is amazing how loud some of these can be too. People who live under flight paths or near railway lines soon stop noticing the roar of the engines as planes take off or the rattle of carriages as trains pass during the night. It becomes routine to us, and so we stop noticing it. And in that, we have one of the first approaches we can harness to get ourselves noticed in terms of sounds.

Contrast

We are hardwired to notice and pay attention to things that are different. Do you remember how we discovered this when we were looking at visual processing (if you pardon the pun)? Well, this is also true for audio information – when something changes in our auditory input it can capture our attention in just the same way. One of the most effective ways for us to harness this is by using silence. In a world where we are constantly bombarded by noises, and voices and music, silence becomes noteworthy. I regularly work with either the radio or a podcast on in the background. I have to admit that often, it blends into the background and I cease to pay attention to it, particularly if I am trying to read, write or do something involving words. However, if there are a few moments of silence, then I am absolutely focused on it. Has the Wi-Fi dropped out? Have we had a power cut? What has gone wrong with the broadcast? If and when this is done intentionally, it is a stroke of genius. This is simply because it then means that, when the content does return, I am inadvertently paying full attention to it. In some ways we can consider this a variation of the 'primacy/recency' effect we have already covered, as it directs our attention and forces us to take notice.

Similarly, sounds that we would not expect to hear in any given context can create almost as good a contrast as silence. Bird song surrounding a business podcast, traffic noises on a radio station that plays relaxing classical music or sneezing on a channel that is dominated by talking can all be powerful methods of contrast. These will pique our curiosity and again trick the RAS into attending to the sounds as the brains of our audience (literally!) try to comprehend the new and unexpected occurrence.

Relevant

Going back to the RAS for a moment, another way that sounds can be used to capture our attention is to select those that are highly relatable to us and relevant for us. We will all have sounds that we are used to, which effectively make up the soundtrack to our lives. But within these will be sounds that occur that we cannot avoid, we just have to respond to.

The sound of smashing glass works well for many environments, as it is often associated with danger, and so we automatically pay significant amounts of attention to its source. For example, in the hospitality sector it means something has gone wrong, and something also probably needs clearing up – quickly. In the home, it often means the same, with equal degrees of urgency if small, mobile children are in the house. However, it can also be associated with a very different sense of danger – that of vandalism or even intruders. Can you see how the use of one sound effect like this can potentially enable you to capture the attention of a wide section of the population? They may all create different, individually relevant, associations to the sound but in each case it will still be compelling and noteworthy for them to attend to.

It is remarkable how discerning we can be regarding these sounds too. I know from experience the way the sounds of children playing in a park can change from shrieks of delight to screams of concern or even pain. These delicate differences can be harnessed to create some very interesting and engaging content across a variety of platforms.

Repetition

The simple repetition of a series of sounds, consistently and over time, creates a strong association with a brand. Jingles such as the 'Intel inside' one, the Ring doorbell chime or even just the Netflix 'da dum' have become some of the most recognized sonic branding devices of our time. With repeated exposure in our lives, these have come to represent a significant amount of the brands' identities, and are as protected as their logos and colour palettes.

Similarly, the current trend for harnessing and promoting ASMR (autonomous sensory meridian response) devices can be used repetitively to great effect. For example, the iconic sounds of pulling the tab off a Coke, snapping a KitKat finger in two or the reassuringly solid thud of a Volkswagen car door can also build brand strength over time, and reinforce their values very effectively.

The key to making repetition work effectively lies in two areas: consistency and memorability. In order to benefit from these, it is important that the same distinctive sound pattern is used over an extended period of time. Too much deviation will reduce the effect significantly and diminish the leverage that is gained over time. Conversely, taking the time and steps required in order to build a solid piece of sonic identity will become a convenient and powerful brand asset that can be optimized for a wide variety of different campaigns and formats.

Infectious emotions

The non-verbal human sounds that we make provide opportunities for marketers to create a compelling soundscape, and one that has some very engaging properties. Some of the sounds we produce have the ability to illicit almost automatic responses within those around us. Consider yawning. I can hardly even type the word 'yawn' without triggering a yawn in myself. How about you? Have you read the word 'yawn' enough now to have created the same effect in you?!

Yawning may not be the desired behaviour we want people to display but other emotions can also be infectious and used more widely to create successful outcomes. Think about the audience laughter that features so richly on TV shows recorded in front of live audiences. This is done to make it more likely that you, the now viewer at home, will also laugh and enjoy the content, and reflect on it in a pleasurable way as a result. More than that, though, when you laugh in social situations (such as when you watch with family or friends), you trigger the release of chemicals within the brain that support the development of long-term relationships.[3] Therefore, this recorded laughter is creating bonds and connections that may well transcend those people in the room with us, and actually make us connect more strongly to the characters within the show too. However, there is a caution to add here. This laughter needs to be genuine. Using so called 'canned laughter' may work in the opposite direction and actually damage the brand and customer relationship, if the audience is able to determine that it is not authentic and sincere.

Many sounds contain emotional connections for us, and this is why they can be used so effectively across many marketing platforms. Music and songs may remind us of key times in our lives, they may make us feel energized, soporific or perhaps just nostalgic. But other sounds can do this for us too. And as the growth of sonic branding continues, these sounds can become highly effective ways to create differentiation as well as convey emotional connections. Indeed, audio assets have been identified as being more than twice as effective as visual assets when it comes to brand attention.[4]

So, using these insights could really be... wait for it... sound advice!

Chapter summary

1. The location of our ears on each side of our head provides us with vital information about the location and source of sounds, as well as their key characteristics.
2. Although most of the sounds we hear are dominated by language, there are three categories of non-verbal sounds that we can consider:
3. Non-verbal human sounds, e.g. laughing, coughing, tutting.
4. Environmental sounds, either artificial (like machinery or traffic noises) or natural (like bird song or crashing waves).
5. Music, which has a myriad of effects on many different aspects of our cognitive processes, physiology and behaviours.
6. The way we access music and sounds has changed significantly. They are no longer likely to be communal or shared experiences, instead we expect to have more choice and control over what we hear (or not!) in any given situation.

7 As with all of our senses, the RAS is responsible for selecting which we may pay conscious attention to. However, we can use methods like contrast, repetition and relevance to improve our chances of reaching that vital threshold.

8 Some of the sounds available to us are infectious and will not just catch our attention but will create enhanced connections between the audience and the brand too.

Notes

1 Matziorinis, A M and Koelsch, S (2022) The promise of music therapy for Alzheimer's disease: A review, *Annals of the New York Academy of Sciences*, 1516 (1), 11–17, DOI: 10.1111/nyas.14864 (archived at https://perma.cc/CCG7-NYXN)

2 Malcman, M, Azar, O H, Shavit, T and Rosenboim, M (2024) How does background music affect dining duration, tips and bill amounts in restaurants? A field experiment, *Behavioral Sciences*, 14 (12), DOI: 10.3390/bs14121188 (archived at https://perma.cc/TBV9-TTPH)

3 Manninen, S et al (2017) Social laughter triggers endogenous opioid release in humans, *Journal of Neuroscience*, 37 (25), 6125–31, DOI: 10.1523/JNEUROSCI.0688-16.2017 (archived at https://perma.cc/J5VS-H3JV)

4 Sheridan, A (2020) The power of you: Why distinctive brand assets are a driving force of creative effectiveness, Ipsos.com, February, www.ipsos.com/sites/default/files/ct/publication/documents/2020-02/Ipsos_Views_Power_of_You.pdf (archived at https://perma.cc/TVK3-53BG)

12

Touch and go

Context

Over recent years, there has been a monumental shift in the way we work, spend our leisure time and definitely in the way we all consume content. That shift has been catapulted by changes in digital technologies. From VR and the Metaverse to software versions of traditional publications, much of our content is now online. But what does that mean for our brains? Is it equally effective for us to access information in this format or do we still need to engage with our physical environment? This is going to be a touchy subject!

As human beings, we have developed and flourished as a result of harnessing the power of our environments. Our ancestors learned to cultivate crops, produce fire, channel water and domesticate animals, all to enhance their lives and provide greater stability and security. Fast forward a few hundred years (well, more than just a few actually), and our more recent ancestors were making furniture, building cathedrals, composing music and writing texts. Their lives were still very tactile, very 'hands on' – unlike the lives we lead today.

You see, according to research by Bionic carried out in 2024, the average UK adult spends 76 per cent of their waking hours online.[1] This translates to 223,015 hours online throughout their career, equal to a staggering 9,292 whole days or 25 years of their working lives. Regardless of how you feel about this statistic, or what emotions that may raise for you, the convenience of digital platforms and formats is undisputed. For us professional marketers and also for us as consumers, they offer unparalleled levels of access, opportunity and cost effectiveness. But is digital technology really the holy grail it professes to be?

Our bodies are designed to provide us with information about our environment. But unlike visual or auditory information, tactile information (based around our sense of touch) requires us to interact directly with our environment – we need to be in physical contact with it in order to gather the information. It is a much more intimate and direct method of learning about our environment, as it does not rely on light or sound waves to convey information. Instead, a network of touch receptors on the surface of our bodies provides detailed information for us regarding the people and objects around us. They inform us not only that contact is taking place, but also the pressure of that contact, its temperature, any relevant vibrations or movements, details of the texture of the item and of course if something is painful to us.

These touch receptors, referred to as mechanoreceptors, convert the information received into electrical signals which can then be transmitted into the brain for processing. Sometimes this involves them travelling the length of our body, as anyone who has ever stood on a piece of Lego, dropped something on their foot or stubbed their toe on a piece of furniture will tell you.

Mechanoreceptors are not evenly distributed across the surface of our bodies. Instead, they are concentrated in the areas where we are most likely to interact with, and explore, our environment. These include areas which I am sure you could confidently identify as being highly sensitive, such as our fingertips, lips and feet. However, one you may be less likely to name is hair follicles. They also contain sense receptors, as they too can be valuable indicators of the presence of objects within our immediate environment, and particularly in the vicinity of our all-important head. Additionally, it is interesting to note that our muscles, tendons and joints also contain mechanoreceptors. These work in a slightly different way though, as they are not involved with information coming in from our external environment. Instead, these mechanoreceptors inform us of the position and location of our limbs *within* it, allowing us to interact with our environment efficiently and effectively.

In addition to these mechanoreceptors, we have thermoreceptors. These, as I am sure you have deduced from the name, provide insights for us regarding the temperature of items we come into physical contact with. These convey messages to the brain with incredible speed, as knowing whether something is hot, warm, cool or cold allows us to behave in appropriate ways, almost immediately. Think about how fast you recoil your hand from touching something that is hot or even drop the item if your brain decides it is too hot to hold. Similarly, we may seek to take evasive action if we place an item of food into our mouth that is too hot for us to manage – rarely a pleasant occurrence for anyone around us to observe!

The information we receive from our touch receptors allows us to interact with our environment but it also conveys important details to our brain. Many of these details can be harnessed by neuromarketers who are in the know, so let's add you to that team shall we? We will do that by exploring some of the ways our brains and decisions can be influenced by our tactile experiences. We are going to start by understanding more about the differences that exist between the way we process written information, which is presented to us either digitally or in hard copy.

Improve function

I am going to give you a choice. I need you to read a large document, say 50 pages long. Would you rather have access to it in digital format or in hard copy?

Now, what if I needed you to edit that document and make amends to it. Would you still choose the same format?

Finally, what if I then needed you to be able to quickly navigate around the document and locate key sections of the text? Would you still choose the same format then?

You see, what you may not be aware of is that your brain responds differently to materials that are provided in a digital format, compared to those that are provided in hard copy. Research published in 2021 examined these differences in detail and concluded that there are many advantages to be gained as a result of providing content in hard copy, printed formats.[2] These advantages include better recall of the content, improved comprehension and it being 'easier' for our brains to access.[3]

Why would that be?

First, remember that the areas of our brains involved in reading text are areas that have evolved from other processes. That means the whole process of reading is going to be challenging for us and deplete our precious neural capacities. We have also discovered that much of the way our brains work is achieved by creating predictions, and using these to shape our expectations and therefore our responses too.

Now, add into the mix the fact that most of what we read on screens tends to be short. Typically, these are made up of text messages, web pages, social media posts and emails. Therefore, we have learned to approach reading on screens from a position of speed and even something we may refer to as 'skimming' rather than reading properly. So, it may not come as a great surprise to you to find out that in comparative terms digital formats are processed faster.

What may be more of a surprise, though, is the fact that we find it significantly harder to engage with content that is presented in a digital format.[4] Just think of what that equates to when we are reading from our screens all day – it is no wonder we are so tired at the end of it! Referred to as the 'screen inferiority effect', it has also been shown that reading longer text from screens is not only harder for us but it also leads to significant reductions in our ability to comprehend and recall the content presented.[5] Additionally, we tend to feel overconfident in our ability to absorb, comprehend and recall information presented on screens, lulling us into a false sense of security that we have understood and will retain what we have read.

At the moment, it is not totally clear why reading from screens is so much more demanding for our mental processes. The most recent theories centre around three key aspects of our reading experience when we have to engage with content digitally instead of from a physical page:

1 **Mental maps:** We know that part of the way we learn things is by creating mental maps (which is part of the reason why mind maps are so useful for recall). Having a sense of where information is

located in relation to other key aspects, such as which side of the page they are located on, whether they are near the top or bottom, etc, seems to enhance our ability to recall it later. However, when we are scrolling down page after page, it is much harder for us to be able to form this internal map, and, consequently, our ability to recall the information is significantly compromised. Additionally, when you have a physical document in your hands, we are constantly more aware of our position within it – are we nearing the end of it or still in the middle? Although digital formats try to provide similar information, we are not as aware of this subconsciously as we are for printed versions, so, again, our ability to navigate and recall is reduced.

2 **Scrolling:** The other element that is believed to contribute to difficulties with processing digital content is the nature of scrolling. A physical page does not move, instead our eyes move over it in regular and predictable ways, and this allows us to access the information on it. Conversely, reading the same page digitally, requires us to keep breaking off, moving the page up and, critically, keeping track of where we are on the page as it moves up. All of this requires much more effort. Even more so if we are reading on our mobile devices, as the volume of text we can easily see on their screens is smaller still, so needs scrolling even more frequently. It is proposed that the resources this process consumes means we have less left with which to understand and make sense of the actual content, hence our comprehension and recall levels being so much lower. We can either keep track of the location of the words or we can process their meaning. Doing both is too much for us, particularly at speed.

3 **Distractions:** When we read a book or a printed document, we may be exposed to a few distractions coming in from the environment around us. People calling us, phones ringing, fights breaking out – you name it, there are many different inputs all competing for our attention, as we have learned. And we have to admit that if the content is not particularly compelling and interesting to us, then we may also have internal distractions vying for our attention, such as

wondering whether it is nearly lunchtime, realizing how cold we are or just needing to stretch and adjust our position for a while. However, with digital formats the potential is vast. Pop-ups, banners, adverts breaking up the content and even notifications from other apps that appear on our screens are all much more likely to derail any attention we have managed to muster and retain.

Now, I realize I may sound like a total Luddite here, and someone who rejects all forms of digital content. I promise you, I do not. I love the convenience of hyperlinks, the release from having to carry numerous documents around with me and of course the environmental benefits of not producing printed material. However, what I would like to suggest is that we do not just deliver everything electronically without first considering whether it is the right format for what we are trying to achieve.

Where and when to use digital

Long pieces of content, by which I mean anything over about 500 words, are usually more effectively delivered into our audiences' brains in hard copy than online. Anything less than that can be equally effectively conveyed in either format, but once you reach that sort of length, then comprehension and learning are both improved if the material is presented in print format.

If you are catering to an audience who may have additional needs, such as changing to an increased font size, a dyslexic friendly font or a different background colour, then again, digital makes this easier. Most devices and platforms allow people to adapt the way content is presented to suit their individual preferences and requirements. This clearly makes it easier for us as marketers, as we do not need to try and find formats and designs that will work for the highest number of prospects or customers. Also, the increase in the use of screen readers and AI-based alternatives that can take digital content and quickly produce something else from them clearly provides advantages that printed media cannot rival.

Then, of course, we need to acknowledge that the sustainability agenda has driven lots of content away from being printed. Forgive my cynicism, but dare I suggest that the 'green agenda' has also been used as a socially acceptable way for us to leverage the other major advantage of presenting content digitally – cost. The savings that can be made, both in terms of printing and distribution costs, are not insignificant, and I am sure are another major factor in explaining why so much of our marketing budgets are now spent on digital communications. Along with these benefits comes the immediacy of the delivery – worldwide. This does provide a high level of advantage as our audiences may become increasingly global, and, in addition, we may want to release access to the same information to all of our territories, at the same time. However, we can now start to see that the two methods are not equally effective, so it is not as simple as selecting based on cost savings or convenience alone.

Finally, as I am sure we have all come to appreciate, presenting content digitally does allow us to have significantly greater levels of analysis and understanding of how our audiences have engaged with our materials. Being able to measure and report on responses, opens and engagement allows us to analyse the outcomes of our work in exciting and informative ways. We get to identify and understand which specific parts of our content have had the greatest impact, lead to the most conversions or created the largest number of referrals. This in turn enables us to build a better understanding of our customers, and make more informed decisions about our marketing strategies and approaches in future. Again, these are good advantages, but *not* at the expense of communicating our message effectively. Why do I say that? Well, read on...

Emotional connection

The final critical aspect for us to consider when debating whether to present content digitally or in printed format is emotions. Yes, they really do pervade into everything don't they?! Don't say I didn't warn you...

Print media appears to create stronger emotional connections and emotional memories than digital formats do. In research carried out by the University of Bangor, Royal Mail and the Millward Brown agency, differences were identified in activity in a number of key areas within the brain.[6] They concluded that the increased activity in both the medial pre-frontal cortex and the cingulate areas (parts of the brain that are integral to our emotional engagement) indicate that the emotions were also internalized more. This would suggest that when viewing physical materials, the participants were more likely to relate the content to their own feeling and experiences too.

These factors combine to create a more enduring legacy when content is viewed in a physical, hard copy manner than when presented digitally. This is achieved not only by the stronger activation of memories and emotional areas within our brain but also the depth of the connections that are made. If we want to create brand awareness and deliver campaigns that have significant impact, digital presentation of content may, for these reasons, not always be the best choice.

However, there are even more reasons, and consequently opportunities, for why our sense of touch can be used to enhance and influence the experience customers have. Subconsciously, we make assessments and judgements about many different aspects based just upon the way something feels when we hold and interact with it.

Assessment of quality

Touch allows us to interact with products in many ways, probably the first of which is when we have the opportunity to test out how it will be to use. From holding the new device in our hands and seeing how it feels to operate to test-driving a new vehicle or even trying on an item of clothing. The creation of these opportunities are a really valuable method for us to adopt if at all possible. Allowing people to test and experience a sample of the full package they will access when they make the purchase engages their sense of touch in more ways than they will appreciate. In addition to that, it will also harness the power of dopamine, and, knowing how compelling anticipation is, it is easy for us to begin to see how we can get our prospects' brains excited. Genuine interaction, which engages as many of their senses as possible, will create a memorable and enduring experience that the brand will become associated with, and desired for.

I am sure we are all very aware of one of the ways we rely on our sense of touch to give us an assessment of quality. This may be conscious, and done as an intentional enquiry (does that fabric crease easily), or it could be subconscious and almost instinctive (running your hand over the smooth surface of a possible new office desk). Are you aware, though, of the consequences that follow as a result of these, sometimes apparently insignificant, gestures? In the nicest possible way, you are probably not... yet. Once I have made you aware, I promise that you will realize how choosing the right board for your brochure, material for your promotional merchandise or even packaging for your product will silently impart lots of information about your organization. These can include the quality and standards to be experienced, the price-point to be expected and even the values to be observed.

Worth the weight

One of the measures our sense of touch conveys to us through the use of mechanoreceptors is an understanding of how much something weighs. Now, I don't mean an accurate measure of its weight (how dull would that make 'Guess the weight of the cake' competitions?). No, I am talking more about relative weight – which of a number of items weighs the most.

The interesting discovery that has been made in this area is that we can subconsciously detect differences in weight between items. Hmmm, interesting... possibly. However, the mind-blowing discovery for us marketers is that we use this information to shape many of our other perceptions too. Research carried out to explore this phenomenon has shown that for products as different as yoghurt[7] and wine,[8] the impression we create of the product is significantly influenced by how much it weighs.

Take yoghurt for example. The same yoghurt was served in three identical looking bowls, which only differed in terms of their weight. Each sample yoghurt was rated by participants for their liking of the product, its density and their price expectations. In each of these cases, the product served in the heaviest bowl was rated as being more dense (rich and creamy) and placed at a higher price point than its lighter counterparts. Yes, you read that right. If you put the same item in heavier packaging, consumers judge its quality to be higher and are prepared to pay more for it.

The same was shown to be true with wines, where the weight of the bottle was shown by participants to be highly associated with both the quality and the price of the wine it contained.

This phenomena can also come into play when we are not making a direct comparison between two (or more) products in front of us. It can also influence our thoughts and behaviours if we just compare the weight we are holding with the expectations that had been created within our brain. Picking up an item or garment that is heavier than we thought it would be will again lead us to make assumptions about its quality and therefore value. If it is an item of jewellery, then heavier will equal more valued and desirable. However, if it is a drinking glass, probably the lighter, more delicate ones will be valued more. So, which would be most desirable for your products or in your sector?

Warmth

There is a very interesting anomaly that has been discovered within our brains. As a result of work carried out in 2008, it was shown that the same area of our brain is implicated in the processing of both physical warmth and also psychological warmth.[9] Big deal you might say! And you would be right – it genuinely really is. Because, you see, one of the first assessments we make about someone is whether or not we trust them. Clearly, it has been vital to us throughout our evolutionary past to be able to quickly determine whether someone we meet should be considered as a friend or a foe. Hence, it is one of the first assessments we make when we meet someone, and we do it within the first 100 milliseconds.[10] Just for reference, that is about the same amount of time it takes for us to blink. Wow.

Now, in the modern business world, we tend to want people to trust us. If we are suggesting a solution for an individual, or a company, we want them to believe us. If we are trying to pitch for some work, we want the panel to have faith in us. If we are trying to get a job, we want to come across as being trustworthy. So, did you know that one of the easiest ways to get people to trust us is to get them to hold a warm drink? I kid you not. Let me explain.

The thermoreceptors in our fingers send a message up to the insula, conveying information about the warmth (or otherwise) of the item we are holding. However, this same area is also implicated in our assessments of whether we feel favourably towards someone. Literally, it influences whether we view them 'warmly' or not. So, by getting someone to hold a warm drink or something when you first meet, you will be helping them to create a more favourable impression of you. Again, that might be interesting but does it really change much for us in the long run? Oh yes, potentially it could.

This warmth is likely to mean you are viewed more favourably, they will be more generous towards you, more patient with you, more tolerant of issues that occur and generally be more charitable towards you. All of which will happen subconsciously but work powerfully in your favour. So, get the kettle on, eh?! And if they do not want a hot drink, get them to hold yours. But whatever you do, do *not* let them hold anything that is cold. This has precisely the opposite effect of making them subconsciously view you coldly, with suspicion and distrust, and will give you an uphill challenge before you even start.

Endowment effect

Another way that we can harness the power of touch is to incorporate a concept referred to as the 'endowment effect'. Simply put, this effect means that if we own something, we are likely to value it more highly than anyone else will. We can see this play out on eBay, Vinted and in auction houses across the country, where current owners set prices or put on reserves that value items higher than a potential buyer does.

There may be a number of reasons for this, but many theories include reference to the fact that we are more motivated by the fear of losing something we have than by gaining something new. Do you remember that from our earlier discoveries? Therefore, it is almost as if we create an invisible connection to items we own. This connection makes us much more reluctant to part with them, and so we seek greater compensation for them when we face that loss. Where this gets even more exciting for us marketers is how quickly this comes into effect.

In their original research carried out into the phenomena, Daniel Ariely and colleagues gave a group of people a mug. They asked them to value it and the median price given was $7.12. Now, a second group of people were only allowed to view the same mug, and asked how much they would be prepared to pay for it. Their median price was $2.87. That is merely 40 per cent of the price the possessors of the mugs wanted to get for them. But remember, these are not their mugs. They have merely been handed them a few minutes earlier. And yet the endowment effect was at work already.[11]

So, we can use this as part of our consideration of touch, to help people value our products and services more right from the outset. Can we give them a sample to hold or try? Free trials of software and subscriptions are a great way to let people experience the benefits for themselves, knowing that they will value it so much more once they have. Even holding a brochure or some related materials can be enough to trigger these sensations of connection and perceived value. I really encourage you to see where and how you can introduce appropriate opportunities into your different customer journeys, as I assure you, they will deliver very tangible results.

Our sensation of touch is often overlooked when crafting and managing the brand experience. In this chapter, we have learned that this is a mistake, as it does in fact bestow a significant number of benefits. We have become very used to navigating our world by placing great importance on what we interact with directly. As children, we explore our world in a very tactile manner, and it is only with adulthood that we become more restrained and conservative in this approach. This is a change that is often suggested or even demanded by society and the world around us, rather than by our internal processes. However, we now realize that these systems are still active within us, and are highly influential in the impressions we form, the intentions we create and the behaviours we manifest.

Chapter summary

1. Our sense of touch, unlike vision and hearing, requires us to be in direct contact with the source of our sensory stimulation.
2. For most of our evolutionary history, we have led very tactile lives and engaged very directly with our environments.
3. Our body is covered with receptors that provide information about the presence, texture and temperature of items around us. However, their distribution is not equal across the surface of our skin – they are concentrated in the most important and likely places we will encounter people and objects.
4. Our brains do not process information presented in a digital format in the same way that they process information presented in a hard copy, print format. As a result of these differences, we find it harder to comprehend and recall information which is presented digitally.
5. Our sense of touch provides us with information which our brain uses to assess quality and make value-based decisions.
6. We experience physiological warmth and psychological warmth in the same area of our brain, giving us an opportunity to create greater connections with people we meet and may want to influence.
7. If we own something, we tend to value it more and seek greater compensation for being parted from it. This effect comes into play very quickly, so providing and using samples can help us to increase the perceived value of our offer.
8. The sense of touch is very underrated within most marketing considerations. If used intentionally, it has great potential to unconsciously influence the perceptions and actions of your target audience.

Notes

1 Couort-Jones, L (2024) How much do we rely on WiFi? Bionic, 17 June, https://bionic.co.uk/business-connectivity/guides/how-much-do-brits-rely-on-wi-fi-2024/ (archived at https://perma.cc/D89H-9584)

2 Venkatraman, V, Dimoka, A, Vo, K and Pavlou, P A (2021) Relative effectiveness of print and digital advertising: A memory perspective, *Journal of Marketing Research*, 58 (5), 827–44, https://doi.org/10.1177/00222437211034438 (archived at https://perma.cc/EJ9T-EJGK)

3 Delgado, P, Vargas, C, Ackerman, R and Salmerón, L (2018) Don't throw away your printed books: A meta-analysis on the effects of reading media on reading comprehension, *Educational Research Review*, 25, 23–38, https://doi.org/10.1016/j.edurev.2018.09.003 (archived at https://perma.cc/Q5KL-JWKN)

4 Clinton, V (2019) Reading from paper compared to screens: A systematic review and meta-analysis, *Journal of Research in Reading*, 42 (2), 288–325, https://doi.org/10.1111/1467-9817.12269 (archived at https://perma.cc/8TTF-33EV)

5 Liao, S et al (2024) Dynamic reading in a digital age: new insights on cognition, *Trends in Cognitive Sciences*, 28 (1), 43–55, DOI: 10.1016/j.tics.2023.08.002 (archived at https://perma.cc/RQ9U-HBFC)

6 Millward Brown (2009) Using neuroscience to understand the role of direct mail, Case Study, https://twosidesna.org/wp-content/uploads/sites/16/2018/05/Using_Neuroscience_To_Understand_The_Role_Of_Direct_Mail.pdf (archived at https://perma.cc/PJ5R-4GG5)

7 Piqueras-Fiszman, B, Harrar, V, Alcaide, J and Spence, C (2011) Does the weight of the dish influence our perception of food? *Food Quality and Preference*, 22 (8), 753–56, https://doi.org/10.1016/j.foodqual.2011.05.009 (archived at https://perma.cc/UUB2-ATTX)

8 Piqueras-Fiszman, B and Spence, C (2012) The weight of the bottle as a possible extrinsic cue with which to estimate the price (and quality) of the wine? Observed correlations, *Food Quality and Preference*, 25 (1), 41–45, https://doi.org/10.1016/j.foodqual.2012.01.001 (archived at https://perma.cc/RFM2-ZEMQ)

9 Williams, L E and Bargh, J A (2008) Experiencing physical warmth promotes interpersonal warmth, *Science*, 322 (5901), 606–07, DOI: 10.1126/science.1162548 (archived at https://perma.cc/LU22-RDAM)

10 Willis, J and Todorov, A (2006) First impressions: Making up your mind after a 100-ms exposure to a face, *Psychological Science*, 17 (7), 592–98, https://doi.org/10.1111/j.1467-9280.2006.01750.x (archived at https://perma.cc/6KGA-2LT4)

11 Ariely, D (2009) *Predictably Irrational*, HarperCollins, New York

13

Smell and taste

Context

Raise your hand if you are thinking 'this chapter isn't going to be relevant to me, we don't have anything scented or edible like that in our organization'. Is that you? Is that what you are thinking? Have you got your metaphorical hand up? Well, let me give you a clue. Nobody should have their hands up at this stage. Seriously. Because all of us need to be aware of the role that these senses play within our daily lives, and, therefore, what they can potentially do for our marketing.

Good. So now we have got you reading on instead of skipping this chapter, let me justify why I make that claim. But before I do that, I am sure you are already aware that your senses of smell and taste are intrinsically linked. Have you ever held your nose to swallow some hideous medicine, so you don't have to endure the taste of it? Or what about when you have a cold and your nose is blocked, rendering you incapable of tasting your food? Yes, you know it already. But do you understand why that is the case? And do you believe that this phenomenon holds very real and significant opportunities for us to create strong and enduring connections with our audiences? No? You will by the end of this chapter, I promise.

Let's start by examining our sense of smell.

Smell

When we experience a sensation of smell, what has actually happened? Well, tiny molecules within the air have been detected by a group of specialized smell neurons (or olfactory sensory neurons if you want to give them their proper name). These neurons are located high up within our nasal passages and are stimulated when they detect scent or 'odorant molecules'. Items in our environment release these odorant molecules, which are then usually brought into contact with our olfactory neurons when we breathe in or sniff. And, just like with colour vision, where some photoreceptors are activated by different wavelengths of light, with smell, different odorant molecules will bind with specific odorant neurons. This binding process will then trigger the release of an electrical signal that is passed to the olfactory bulb beneath the frontal lobe of our brain. The good news is that this area is once again very close to the top of our nasal passage, so the signals are very quick to be received there. The final step in the process is for the information to be conveyed to key parts of our brain such as the amygdala and the hippocampus to enable us to identify what the smell is.

This last stage provides us with the first interesting part for you to know. All the other senses relay their information to the cerebral cortex via the thalamus and the RAS (Reticular Activating System – I am sure you remember that by now), where it gets processed and filtered. However, smell does not. Olfactory information bypasses this stage. This means it goes direct to the limbic system, which, as we know, houses our core emotion and memory functions. For this reason, our sense of smell is very powerfully and intensely connected to both our memories and our emotions.

It is thought that this process relates back to our animal ancestors who would have been much more reliant on smell than we are today. Primitive vertebrates not only used but heavily relied on their ability to detect, identify and locate a large range of smells in order to facilitate their survival. Please don't think I am exaggerating when I say that, as it would literally have determined their ability to detect sources of food, mate successfully and also keep them safe from predators too.

Although times have changed, and we are no longer so reliant on smell, the legacy of these processes continue to be highly effective and influential within our brains today.

Have you ever smelt a scent that just hits you with a wave of emotion or memories? Maybe it is the perfume your grandmother used to wear or the smell of the hot seats inside a car your parents used to have when you were a child. It could be the smell of a steam engine chuffing past, a particular meal being prepared or even a distinctive plant or flower. Because the processing aspect that usually takes place within the thalamus is bypassed, we can find ourselves experiencing very intense emotions and memories far faster than we might from our other senses. One 'whiff' and we are there again, in the back of our childhood car or standing on the platform. We are able to recall tremendous details and events, and we are feeling all the emotions alongside them too. However, it may take a while for our more conscious processes, such as language, to be able to catch up with this fast-track response. Therefore, we often find ourselves recognizing a smell, being able to state whether it is a good association or a bad one but not actually being able to label or articulate it sensibly. Can you recall times when that has happened to you?

One of my favourite smells that triggers strong emotions and memories is quite recognizable to me – it is the smell of Polo mints. These reassure me that everything is going to be OK, they make me feel very safe and very loved. Polos! I know, it is strange, isn't it? But hear me out on this one. You see, my father always carries a tube of Polo mints in his pocket. Always. And he will usually have a handkerchief in there too. So… as a child, if I fell over or hurt myself or was upset for any reason, Dad would produce a handkerchief out of his pocket for me. And what would it smell of? You've guessed it. Polos. Hence the association I have with them to this day. Thanks, Dad!

Before you judge me, take a moment to reflect on what smell associations you might have! Remember, they can be positive (as in my example above) or negative (like the smell of whatever that hand soap was that they used to put in the dispensers in primary school – yuck!). See if you can come up with two or three smells for both positive and negative categories that hold strong emotional connections for you. Are they recent or have they stood the test of time from your childhood? Are they easy to describe and define or are they made up of a mixture of stimuli (like the smell of an old wooden cupboard at my grandparents' house, which was made up of a varnish smell, the slightly musty smell of furniture that may be a bit damp and the smell of the lavender shelf liners they used to use)? Although not easy to define, it is distinctive to me, and if I smelt it again now, I would recognize it in a heartbeat.

The interesting thing about getting you to do this is that I bet you were almost conjuring up the sense of those smells while you recalled them too weren't you? It is a hard process to do, particularly when you compare it to what we can achieve with our visual or auditory imagination. However, most of us at least try to recall smells when challenged like this, even though it is generally far less effective! So, if you had a go, well done you for trying! I am proud of you.

How brands are utilizing scents

So far, we have learned that smells create very strong emotional connections, and trigger our memories very quickly too. It is no wonder then that many brands are catching on to the potential of this information. Supermarkets have known (and exploited) for a long time the fact that smells like freshly baked bread stimulate our appetite, making food seem more appealing to us, so we therefore spend more. It does not just stimulate our appetite though: that smell in particular conjures up associations of a very wholesome and homely nature, which will affect the way we feel about the food purchases we are about to make. We are likely to view them more positively, feel less guilty about the volume and variety of food we buy, and also be inclined to judge them as being more beneficial for us than they actually are. And now retail stores like Abercrombie & Fitch have developed their own fragrances to control the smell sensation their customers experience when they cross the threshold into their stores. Some of these fragrances are added to product lines, whilst others are not for sale, they are just used in store to control and manage the customer's experience and, dare I say it, mood. Like the infamous 'new car smell' that we all recognize and love so much. Do you realize that this is an artificial construct? It is entirely man-made, and has nothing to do with the new interior fabric, paintwork or the smell of a new engine bedding in.

It is worth discussing here that there is a debate to be had around whether scents need to be consciously detectable or not in order to be effective. Some of those smells identified above would be strong enough to be something we notice and are therefore consciously aware of. Indeed, you need this to be the case if you are going to offer the scent for sale – it needs to be one that people not only notice but then, in turn, desire. However, this does not need to be the case. Just as with our other senses, we process significant amounts of olfactory information below our conscious threshold. This means that many of the scents that brands and organizations are using will be much more subtle than these. They will be smells that we process unconsciously, so never reach our conscious awareness threshold. They may therefore be categorized as being subtle, yes, but still highly effective.

The reason some brands are doing this is two-fold. First, it allows them to cover up any other potential smells that may not be received positively within the audience's brain. Take the smell of a retail clothes shop for example. If no additional smells were added, the residing odour you would probably detect would be one reminiscent of plastic and cardboard. The packaging that clothes are delivered in will dominate the smell that the fabrics adopt. For some people this might be acceptable, but for others it might have more negative associations, and even feel artificial and cheap. These responses are unlikely to be consciously detectable unless they are really severe, but as you have learned, their unconscious detection can be damaging enough. Better, therefore, to mask this odour by introducing one of your own that will minimize, or even totally cover, the 'naturally' occurring scents created by the processes of manufacture, storage and delivery.

The added advantage of this is it also allows the organization to adopt the second reason for creating your own smell as some brands do – to create consistency and reinforce brand recognition. In doing so they completely manage the odour that exists within the shop, and ensure no differences occur as a result of reactions with any factors outside of their control (as perfumes do when worn by a human body). They literally create their own 'essence'. Therefore, all shops, in all cities and in all climates around the world, will reinforce the same brand memory. So too will any products that are bought online, as we can ensure the same scent is incorporated into our tissue-paper or packaging, which releases when the delivery is opened and the item unboxed. Clever eh? The store really does get brought to you.

Hold on though, it gets even better. You see, as I hinted at above, the 'right' scent can change a buyer's behaviour. It can encourage them into a shop, encourage them to stay longer, to not notice how long they stay and to feel more favourably towards the items they see whilst there.[1] All as a result of the scent they are experiencing. And please do not make the mistake of thinking that we are talking about minor changes here. For example, in one early study into the effects of scent on consumers, it was found that the 'right' ambient smell in a casino meant people put an average of 45 per cent more into the slot machines.[2] Forty-five per cent! Can you imagine what a result like that would do to your profit margins? Initially, in this case, the scents were introduced to just mask the smell of cigarettes, produced by the many gamblers who smoked as they nervously whiled away the hours at the slot machines. However, for obvious reasons, the use of artificial scents has continued, expanded and since become a very lucrative addition for some organizations and brands.

How can service-based businesses utilize scent?

These phenomena are not exclusive to retail and hospitality though. In fact, there is a case to suggest that some of the best scent branding results can be achieved for service-based businesses. If you think about it, it is much harder for our brains to make comparisons, reach conclusions and ultimately make effective decisions, when we have nothing tangible to work with. As marketers, many of us may be aware of the challenges this presents. It is harder to create contrast and demonstrate differentiation. Typically, we rely on adapting our messages and our approaches to convey more specific aspects of our offer, for example the people, their experience, our values, etc. However, in true neuromarketing style, what if we consider this same challenge from the perspective of our prospects? We have learned how difficult it is for their overwhelmed brains to make decisions at the best of times, let alone with less 'definite' information to work with. What we discover is their decisions will become based on other elements like trust, confidence, reputation and the faith they have in the person fronting the business. The good news is that smell can facilitate our creation of each of these. Using an intentional and relevant scent in the reception area of our legal firm, in the printed materials we send out from our architectural practice, near our desks in the travel agency or on our exhibition stand at the industry conference can shift the perception people have towards us completely.

The final point to make regarding the potential for harnessing our sense of smell is this idea that we need to be intentional about how we use it. Many of the items we produce will have an odour of their own, naturally. Think about the smell when you first cut into a box of brochures, walk into your showroom or open a delivery of promotional merchandise. It may be pleasant or aversive, natural or artificial, delicate or overpowering. However, the chances are some form of smell will be there. Scent molecules will be released, and we will be unconsciously affected by them. So, let's start to raise these above our conscious threshold. Make a real effort to become aware of these unintentional scents, and consider the effect they could be having on your brand. Or are you happy to just remain 'nose-blind' to them?

Taste

And so we move on to consider the sense of taste. This is definitely the most niche of the topics we will explore, as there is a limit to the number of sectors and organizations who can utilize this aspect. For most of us, the options may be more limited, but I would suggest they are still not nil.

So, let's begin by learning how taste works. Similar to our experience of scent, taste is based upon receiving tiny molecules but this time from our food (or whatever else we might be tasting). These molecules interact with our saliva, and through the process of chewing, come into contact with the taste receptors we have located on our tongue, and within our mouths. These taste receptors (or taste buds as we might more commonly know them) are activated by different types of chemical molecules, and then convert that signal into electrical impulses. I am sure you are used to this routine by now, but these electrical signals are then sent to the brain for processing, which informs us of the taste sensation we are experiencing. The main tastes can be divided into five distinct categories:

1 **Sweet** These are usually activated by the presence of sugar or derivatives of it such as fructose.
2 **Sour** These are typically acidic chemicals that contain hydrogen ions, such as lemon juice.
3 **Salty** Produced by either salt crystals made from sodium and chloride or from mineral salts which contain elements such as magnesium or potassium.
4 **Bitter** Probably our most vital taste in terms of our evolutionary survival, as it often indicates whether something is poisonous or not.
5 **Umami** Sometimes referred to as 'savoury', this flavour is quite 'meaty' and is usually caused by the presence of glutamic or aspartic acids.

Before we move on from this section, I want to just dispel a myth that you may believe or even be perpetuating. This is the erroneous view that different parts of our tongue are made up of zones that are dedicated to respond to specific tastes. This is not the case. Instead, we have learned that almost all parts of the tongue can sense all of the five main flavours, with two exceptions. First, our tongues contain a higher concentration of all taste receptors around the sides and edges of our tongue than is located within the middle. So there is less density of taste buds in the centre of our tongue than there are around the sides, at the front and towards the back. Secondly, there is a slightly higher coverage of taste receptors dedicated to experiencing bitter, at the back of our tongues. We are not totally sure why this is, but it could be explained by reference to the survival concept outlined above. The concentration is higher at the back of our tongue as it could be potentially vital for us to have a 'fail safe' in place before we swallow any foods, to ensure they are not likely to do us harm in any way.

Interestingly, our ability to detect and differentiate flavours is established within the womb.[3] It appears as though we are born with a natural preference for the sweet and umami flavours. These will usually draw us towards foods that are high in both calories and protein, therefore being vital for our early survival and development. Conversely, we respond negatively towards the bitter and sour flavours from the outset, which naturally leads us away from foods that may be harmful or poisonous to us. And well done if you are astute enough to realize that this only tells you about four out of the five flavours we can detect – nothing gets past you does it? Regarding the fifth flavour – salty – our ability to detect this does not develop until we are about five months old.

Nature, therefore, plays a large role in our taste preferences. But so does nurture. You see the foods a mother eats whilst pregnant affect the make-up and flavours of the amniotic fluid that encases the foetus. Similarly, her breast milk will pass on flavours to the baby once it is born. These, along with the feelings of security and comfort that are usually also conveyed during feeding times, will go on to create food and flavour preferences as the child develops and grows up.

The role of the nose in taste

Now, let us move on to explore the role of our nose when it comes to taste – a process that is technically referred to as retronasal olfaction. Clearly the five taste categories outlined are a long way from creating the vast array of elaborate flavours we can enjoy and identify from our foods. The difference is made up from information that comes from our nasal passage, and the olfactory receptors we have learned are located there. You see, when we chew food, molecules of scent get released, which then travel from our mouths, up the back of our throat and to our nasal cavity where the receptors are located. As I am sure you have discovered many times, albeit inadvertently, this passage is very useful to us when we sing or talk. However, it is less useful to us if we try to keep a mouthful of liquid in whilst laughing or coughing, as it often forces the liquid to come down our nose instead. Either way, we can appreciate that our mouth and nose are connected by a shared passage, formally labelled the pharynx. When the brain receives information from both your olfactory receptors and the taste receptors (buds) within our mouth and throat, the information is combined to give us the experience of flavour. Ta da! Lots more flavours and a vastly richer eating experience has now been unlocked.

Which is also why having a blocked nose impairs our experience of flavours, as we are then reduced back down to just the five tastes that our taste receptors can identify. We might be able to tell that what we are eating is sweet, but we will not be able to accurately report whether it is melted chocolate, honey or maple syrup, for example. Another aspect that might help us, though, is the additional information we receive from our mouth regarding the texture of what we are eating.

Within our mouths, we have a high concentration of mechanoreceptors, or touch receptors if you were not paying attention in the previous chapter on touch! These provide vital information to our brain about the temperature, texture and shape of the food in our mouth. Through these insights, our brain will be able to tell the difference in consistency between the three examples I provided in the above paragraph, and of course, many more combinations besides. Also, the mechanical processes of chewing and manipulating the food within our mouths gives us additional information about how 'chewy' or how hard it is, for instance.

'Thank you very much Katie, this is all very interesting. But how does it help me?'

Good point. Let's get back to what this means for us marketers, and how we can utilize this knowledge. First, as I am sure you are already anticipating, there is great potential for us to harness as much as we can from our understanding of the relationships between food and mood. You probably do not need me to tell you that what we consume affects our mood. I can do that in just two words – 'caffeine' and 'camomile'! But many of the influences are much more delicate and hard to detect than those two, so let's look at a few here.

Health and wellness

Recent trends have emerged around the development and promotion of food products that overtly claim to support our moods. In 2024, Unilever launched a range of three Magnum ice-creams, labelled 'Euphoria', 'Wonder' and 'Chill'. They use flavours such as fruits and chocolate, alongside textures like nuts, biscuit or even popping candy, all encrusted within their iconic chocolate shell, to create very different taste experiences.

So would these flavours actually change our mood? Well, in theory, yes, they could. Flavours like vanilla tend to make people feel very calm and at ease, often evoking nostalgic associations with childhood and home-baking. Conversely, flavours with a citrus base such as lemon or orange tend to be more energizing and invigorating for us, as their 'refreshing' tones dominate. But what about something like peppermint? That may make you think of sweet delicacies that get produced at Christmastime or possibly of medicines and highly clinical applications. And that is the key thing here. Flavours absolutely can, and do, affect our emotions. But most of these effects are highly dependent on cultural sensitivities and individual experiences too. Therefore, they are not easy to deliver on a global scale.

So, how can we approach the area of taste within our marketing strategies and plans? Well, there are essentially two considerations we need to explore:

1. *If you produce food or edible products*

To begin with the easy one, if you are involved in marketing food or edible products then clearly there is a need for you to understand the way your audience feels about the flavours and products you are creating. We know that using physiological measures is going to be more reliable than asking them to give feedback on what they have tasted, and so maybe we can use neuromarketing to establish and analyse their true reactions. We can also use these to test modifications to well-established recipes (such as reducing the salt content or substituting plant-based alternatives) to ensure it will still be well received by your loyal customer base.

However, we need to appreciate that the results we witness will be in response to the combination of factors outlined. Therefore, different samples may need to be provided that either control or explore the variations of texture and temperature, as well as the flavours being offered. Without that, we cannot be sure that we have truly captured the response to the isolated variable we are intending.

Neuromarketing has also been used to test the way food is presented – the packaging, language used to describe it, dishes it is served with, etc – as well as the effect it has on us physiologically and emotionally. There are clearly some very significant implications for our health, societies and overall longevity contained within much of this work too. As many of us struggle to maintain a healthy size and weight, and our healthcare providers are becoming less and less able to cope with the implications of this, food producers are under increasing pressure to play a significant role. Neuromarketing uniquely helps them to balance the conflicting needs of different stakeholders, and achieve truly sustainable outcomes.

2. *If you do not produce food or edible products*

But what if you are not in that industry yourselves? Does this mean you have nothing to learn from the last few pages? Absolutely not. You too can become more considered and intentional about the ways you incorporate taste into your communications mix. Everything from the coffee you serve to visitors to the sweets you give away at events should be approached with new eyes and new insights. Even the taste of the sticky part of the envelope when we lick it to seal it has the potential to affect the way our brand is viewed and remembered. Just imagine if a stationery company were to produce envelopes with flavoured seals. Special festive flavoured ones for Christmas cards, cocktail-inspired ones for happy 30th birthday cards, cake-flavoured ones for children's birthdays, etc... I know I would be up for trying some of those out, wouldn't you?!

As with any marketing investment, the decisions should be driven by the way you want the audience to react, feel or behave. Do you want it want to create associations that are fun, novel, nostalgic, safe, luxurious, innovative, practical, etc? Knowing the answer to this will give you some direction for the categories of flavours that may be most relevant. It will help you to select whether branded jelly beans will be a good idea or more sophisticated mints for example.

Through understanding and appreciating the origins and processes of our smell and taste senses, we can see that there is a lot of potential available to us. Instead of assuming it is for other organizations or other sectors to explore, I hope you can now see that they should be a consideration for us all. Their direct influence on our emotions, mood and memories is too powerful for us to neglect.

Chapter summary

1 Despite what we may initially believe, consideration of the senses of smell and taste should form part of all of our marketing strategies.

2 Smell bypasses the first stage of processing within the thalamus, so it has a fast-track connection direct to the area of our brain responsible for our emotions and memories.

3 Just as with the other senses, smells do not need to be consciously detectable in order to influence our moods, decisions and behaviours.

4 Brands can introduce smells to mask existing scents, provide consistency and even create brand associations and new product lines.

5 Service-based businesses also have significant potential for leveraging the benefit of managing the scents that are within their control… arguably even more so.

6 Taste receptors on our tongue provide basic information about the five key tastes – sweet, sour, salty, bitter and umami. In order to get richer information about the wide array of flavours we can detect, we need to incorporate information from our olfactory receptors too.

7 Different flavours have the potential to change our mood, energy levels and behaviours, to say nothing of the associations and memories they may trigger at an individual level too.

8 Whether we produce edible products or not, we should all explore and intentionally consider the role that flavours and tastes can bring to our customer journeys.

Notes

1 Rimkute, J, Moraes, C and Ferreira, C (2015) The effect of scent on consumer behaviour, *International Journal of Consumer Studies*, 40, https://dx.doi.org/10.1111/ijcs.12206 (archived at https://perma.cc/W44U-UXAW)

2 Hirsch, A R (1995) Effects of ambient odors on slot-machine usage in a Las Vegas casino, *Psychology & Marketing*, 12, 585–94, https://doi.org/10.1002/mar.4220120703 (archived at https://perma.cc/WBV8-AG6R)

3 Ventura, A K and Worobey, J (2013) Early influences on the development of food preferences, *Current Biology*, 23 (9), R401–R408, https://doi.org/10.1016/j.cub.2013.02.037 (archived at https://perma.cc/DAR9-7LH5)

14

Pick 'n' mix

Context

Well done. You have spent a lot of time learning about each of the senses and the ways they are received and then processed within the brain. You have also learned some tips and techniques that you can adopt in order to capitalize on this information. However, you also need to accept that your brain does not work like this in real life. You do not just receive information coming in from one of our senses. Instead, you receive information from a combination of your senses, and you use that to build a more reliable awareness of what is around you. Or so you might like to think...

Growing up, I lived in a house where music was a major factor. In fact, it still is in my home today but that is a different story. The reason I am telling you about my childhood home is because I remember being exposed to a very strange thought experiment there: which would you rather lose... your sight or your hearing?

Now, because our house was often filled with people who are very musical, there was a debate to be had. Could they contemplate a life without music in it... or, perhaps more accurately, could they contemplate a life with only the music they already have inside their heads? What would that be like? Better or worse than a life without sight? Never seeing your children's faces again, witnessing another spectacular sunset, etc. I promise you, I had a great, and very happy childhood; please do not worry that every day contained such bleak deliberations!

It proves my point though. For those of us who are gifted with all five senses, we may not give them equal priority. We may consider some to be 'more important' than others. We may depend on some more than others. We may enjoy some more than others.

One of the strange effects some people experienced as a result of contracting the Covid-19 virus was the loss of their sense of taste. This may be something that we have all experienced at some point in our lives as a result of having a cold. But in the example of having a cold, this loss is usually temporary, only short-term and lasts no more than a day or two. With Covid though, the effects sometimes lingered. Indefinitely. According to research published in 2023, 60 per cent of people who were infected with the virus that caused Covid in 2021 lost some taste or smell functionality.[1] Sadly, only three-quarters of these made a full recovery, with almost 4 per cent never getting their taste or smell sensations back.

How devastating would that be for you? I mean reading that statistic just now, how do you feel? Honestly. Come on, we are friends now. Do you feel that it would be devastating to never be able to taste or smell anything ever again? Or do you feel that actually that would not really be such a big deal? Well, a study carried out in 2022 suggests that you feel it may not be very devastating at all. In fact, some people would happily 'sell' their sense of smell in order to keep access to their mobile phones or even their hair.[2] However, the reality of the situation may be quite different. In real terms, going through life without the ability to taste or smell things does affect us. It makes meals boring, it deprives us of some of the richness in our lives and it also is potentially dangerous as we lose two of the mechanisms we have evolved to help ensure our ongoing survival. So, be careful what you think you would be prepared to go without!

Understanding the relationship between the senses

When we look at the relationship between our senses, we can see that over the years a number of attempts have been made to establish a robust hierarchy within them, with differing degrees of success. One reason for this is the deceptively simple issue of trying to determine what qualifies for any one of the senses to reach the top of the list. What would it take for them to be considered higher or better than any other sense? What are we measuring or defining success by?

Aristotle, the Ancient Greek philosopher, is credited with first proposing that the senses should be placed in the following order of priority:

1 sight
2 hearing
3 smell
4 taste
5 touch

His arguments for why they appeared in this order were based on our continued survival, and priority was allocated to the sensory experiences that would sustain us for longest. Sight and hearing can present us with vital information about things that are a significant distance away from us. They therefore give us opportunities to change our direction or speed, or to prepare in some way, before a close encounter takes place. Whereas, as we have already noted, for us to experience taste and touch we need to be in direct contact with the item. The intimacy of the stimuli means we have less choices available to us. Smell is clearly located in the middle ground, as it falls between those two categories, providing information about something that is close to us but that we do not need to be in direct contact with. However, in the 2,400-odd years since Aristotle was alive, neuroscience has allowed us to take a more scientific approach to the way our senses are prioritized within the brain.

If we look at the relative size of the key areas devoted to processing sensory information within the brain, does that tell us anything? For example, processing visual information takes up approximately 30 per cent of the cortex, whereas the area where smells are processed is more like 0.1 per cent of the brain for us humans. So, does that give us evidence to challenge, change or even refute Aristotle's assertions?

Actually, please do not put too much effort into answering that – it is a bit of a trick question because the priority the brain gives to information coming in from our senses varies. There is no, one, predetermined order that is always applied. Instead, the brain uses a range of processes to determine, in any given situation, which sense should take the lead. I am not sure how you feel this process may take place, but let me clear up one thing before we really start exploring this topic. This is not dependent on, or driven by, the allocation of those precious resources you have heard me refer to on many occasions within this book. Instead, this is examining which sense dominates if information from two or more senses are competing or potentially conflicting.

Priming

Have you seen some of those trends that crop up on social media where a short audio phrase is repeated over and over again? However, as you listen to the phrase being repeated, you are encouraged to read two (or more) different pieces of text? What you hear the phrase saying will change depending on the text you are reading at that time. In one example of this, the two phrases you can read, are 'green needle' and 'brainstorm'. Not exactly similar you might think. However, when you hear the audio track, it genuinely sounds like it is saying 'green needle' when you read that, and the same piece of audio sounds like it is saying 'brainstorm' the moment you switch your eyes to read that text. No matter how you try to change from reading one text to the other, switching back and forwards as if trying to catch out a non-existent pattern, the sound always appears to match the text you are reading at that time. What we believe we hear has changed because we are being primed to hear that, as a result of the words that we are reading.

Let me give you another example. In 2001, a now infamous study was conducted that sought to understand the role of language in the perception of wines.[3] The researchers asked 54 wine experts to evaluate two glasses of wine: one red and one white. The experts described the red wines using terms that are typically associated with red wines, such as 'full-bodied', 'jammy' and 'robust'. Meanwhile, the white wines were described using terms that are more commonly associated with describing white wines, such as 'light', 'crisp' and 'zesty'. However, as I am sure you have guessed already, this was no straight-up wine tasting. The wines the experts tasted were actually both the same white wine. The 'red' wine was merely the white wine with flavourless food colouring added to it to give the appearance of a red wine. And that priming was enough to change the way the wine experts experienced their wine and, consequently, described it. Experts, remember!

Clearly, in this situation, the information coming in from their eyes was in control. That drove the brain's expectations and predictions, which then skewed the 'sensations' experienced by the other senses. They say we eat with our eyes, and we literally do! When we were exploring the way our taste receptors work, and the additional information that is conveyed to our brains by our mouths (such as texture etc), we could also have discussed this there. Because much of the information about texture, flavour, etc will already be 'set' in our brain by the way the item looks, feels as we pick it up, responds to being on the fork, etc. All of this information helps us to piece together our overall experience of eating something. However, we must accept that, as in the experiment cited above, this leaves us vulnerable and open to making significant mistakes. Maybe we think we see something move as we walk down a badly lit corridor just because we have recently been listening to stories of haunted houses. Or perhaps we convince ourselves that the water is freezing cold just because we have been told it is. These are examples of where our senses overlap and where priority is given to one, which then creates an expectation about what will be received in another.

Does this really matter? Well, in some instances, yes, it really could.

At the low end of the scale, the ability to control people's sensory attention is harmless and even entertaining. For example, they are the devices that magicians use to distract our attention and divert our senses away from the important changes that are going on in the act. But we do not have to move far from here before we see how the same approaches can be used by people who pick your pockets, spike your drinks or try to scam you out of your hard-earned savings.

These differences can also explain why a group of individuals who witness the same situation will report it differently. It does not mean that one of them is lying or covering something up, it simply means that their Reticular Activating System (RAS) is focused on different priorities, so different aspects of the incoming sensory signals were allowed to get through to their conscious awareness.

Perhaps more significant and interesting to us is the fact that these biases can affect our internal processes too. Think about what happens when we try to multi-task. Maybe we are trying to eat our dinner in front of a good film on TV. When we do this, our taste experience is reduced, as we are paying more attention to the characters and plot on the screen in front of us. We therefore find the meal less enjoyable and eat quite 'mindlessly', which often means we end up eating more too. But what about if, instead of eating we were trying to talk to someone. And instead of watching TV we were trying to drive a car. Then can you see where problems may arise? Our brain could prioritize the conversation, our attempt to change the podcast we are listening to, the bag of sweets we are trying to open or the children fighting in the back seat over the signals that are coming in from our senses about activity on the road.

We know that this affects our memory, and our attention, but it also affects our decision-making. The real problem here is that some of these changes can be quite temporary, and change a lot over time. This, unsurprisingly, makes it even harder for us to predict and try to mitigate against these in our materials. So what options do we have? How can we even attempt to counterbalance any of these sensory biases?

The first step is for us to really appreciate the different processes that are at work within our brains. How does the brain decide when, how and what to prioritize? The main methods that we are aware of so far are as follows.

1. Context

The context that we are in will be one of the key elements that helps the brain to determine which sense to favour at any given time. For example, when we are in a conversation with someone, particularly if that is taking place over the phone, then we will prioritize audio information. We are essentially tuning in to the information that our ears are providing and dialling down what comes in from the other senses. Conversely, if we are navigating our way along a crowded street or driving down a busy road we will prioritize visual information.

Have you experienced situations where there has been a change in the priority within any given context? I am sure you have. Maybe, as in the example above, you are on the phone to someone and you are giving the audio information priority. However, your partner comes into the room and is frantically trying to signal something to you. At this point, the priority allocated to audio may reduce, and you will instead try to translate the gestures and signals that you are seeing visually so you can respond. During this time, the attention and detailed information you are gathering from the audio input (phone call) will deteriorate significantly. It will soon return, though, once you have made sense of the interruption and responded accordingly, or, alternatively, let them know that you are not giving them the attention or response they are craving at this time.

2. Experience

As we grow older, and we have accumulated more experiences within life, our brains will use that information to evaluate and determine which of the sensory information we are receiving is likely to be the most relevant at that time. In a way, this builds on from what we learned about context above but this time it is also drawing on what we have experienced before. If we know that a particular uncle makes the most amazing homemade pasta sauces, then when we sit down to eat at his table we may choose to focus on the information we are receiving from our taste receptors above all else. This is not required by the environment or necessary for the task of eating, it is a choice we make based on our previous experience.

3. Relevance

If we are trying to complete a specific task that requires intense involvement of one of our senses, then the others will be 'dialled down' in order to allow it to dominate our conscious awareness. For example, if we are trying to finish reading a section of text, be that a report, book, letter, etc, then it is likely that information from our eyes and visual system will be in pole position. Anyone who is trying to catch our attention is going to have to do something significant to make us hear them or become aware of what they are doing. This can be confused with the 'context' aspect but in this category it is all about the task. It is not the location or environmental aspects that are behind the priorities, it is the task that we are attempting to complete that determines them.

A great example of the difference here is illustrated by one of my uncles. I have already told you that I am from a very musical family, and my father's eldest brother was possibly one of the most musical of them all. I remember as a child being fascinated, dumbstruck, impressed and appalled in equal measure by his conduct one Sunday lunchtime. You see, such was his passion (and knowledge) for music that he would often commence humming an opera or concerto, and he would not deviate from his task until he had completed the whole thing. Usually, this may just be a little strange but not dangerous. However, on this particular day, the timing was out. He was just building up to the crescendo, the big finale of this particular piece of work, right at the moment he was being asked to carve the Sunday roast. The 'context' theory would have had him being mindful of the carving knife and fork in his hands, and the people waiting eagerly around the table to eat. However, the 'relevance' part was clearly in charge at that time and, as he only had a few bars of the music left to complete, he did so. Using the carving knife and fork as a conductor would use their baton. Terrifying? Impressive? Disturbing? You decide! But whatever conclusion you come to, it is definitely a good example of how sometimes what we are doing can be deemed more important than where we are!

4. Emotions

It may come as no surprise to you by now that our emotions also play a big role in the way our senses interact. This is a real opportunity for us because it appears that, regardless of the specific sense the content is presented to, something that is intensely emotional will usually dominate our conscious attention. Seeing or hearing someone laughing hysterically, or sobbing intensely, will be given priority over the other pieces of information vying for your attention. This is therefore a bit of a hack that we may be able to use if we are not sure of any other options.

5. *Combination*

In some instances, the brain will use information from the different senses to help it interpret and understand what it is being presented with. This strategy pulls together information, apparently simultaneously, to make sense (pun intended!) of sometimes conflicting incoming signals. For example, when viewing some optical illusions our instinct is to reach out and touch them, to help us process the fact they are actually two dimensional. Or an example I witnessed a child do just the other day: they were eating something and had come across a section of it which I assume had a distinctive flavour or texture. They appeared to be unable to accurately identify it, so they took the food out of their mouth, briefly examined it using their eyes and on recognition, put it back in again and carried on eating! Not very pleasant to watch, and also not a great display of table manners, but another good example of how we can use combinations to help us make sense of things that are going on around us and to us.

A study carried out in 2025 made two important discoveries relating to this phenomenon of combining the information that comes in from our different senses.[4] First, they identified that there are some neural systems deep within our brain that can be shared by different sensory inputs. Regardless of whether the information comes in to our olfactory lobe or our auditory one, they can apparently still share some of the same systems deep in the centre of our brain. But, more interestingly, they also identified that all sensory input, regardless of which sense it came from, activates two of these deep regions: when we are focusing or concentrating. Although there is more to understand about this area at the moment, it does suggest that there are greater areas of overlap than we may previously have believed, understood or anticipated.

6. Space and time

Building on from this, the brain appears to take into account the location and timing of information that is received from the senses simultaneously. If they occur at the same time, such as seeing a flash and hearing a loud bang, then this combination will become a higher priority than other information we may have previously been attending to. Similarly, if a loud crash and a strong smell both appear to come from behind your teenager's door, then you may find that commands much of your conscious resource too.

Joining up the timings and/or locations of incoming information shows that the brain is processing these inputs simultaneously but not totally independently. There is an area that integrates this information and uses it to form a coherent impression of the situation and circumstances around us. So, can we use this knowledge to help us get noticed and processed with more conscious resource?

7. Compensation

It is worth mentioning here that the preferences and priorities we may usually adopt can easily change as a result of injury or training too. For example, a few years ago my daughter, who was nine years old at the time, decided to be blindfolded for 24 hours to raise money for the Guide Dogs charity. Even within that relatively short amount of time, her other senses began to step up and compensate for the loss of information from her visual system. This ability our brain has to divert resources, change priorities and adapt to different circumstances means people can be trained to be piano tuners, art appraisers, flavour creators, neurosurgeons and perfumers just by developing their particular skills in each of these domains. At the less positive end of the scale, people who lose some of their sensory abilities through either accident, illness or injury can also find that their other senses become heightened and refined, providing greater degrees of depth and insight than they previously might have done.

8. Cultural dependencies

Finally, there seem to be differences that emerge in terms of the hierarchy of our sensory information, based upon the culture we belong to. Research carried out in 2018 took this one step further by exploring the relationship between language and sensory hierarchies. Having studied 20 diverse cultures from across much of the world, they found that there is no universal hierarchy that exists.[5] As we discovered in Chapter 10, the words we use to describe the world around us affects our perception of it. Therefore, studying the approach different languages take towards describing and categorizing items within their different senses is thought to give some indication of the likely perceptual processes and priorities they experience.

Essentially, the differing inputs from all of our senses are managed by the brain in a very dynamic and complex manner. Internal and external factors will affect the way the incoming information is processed, and the same person will do it differently on different occasions and under different conditions. This is to say nothing of how different people will approach the same array of sensory information.

As frustrating as this might be for us marketers, this ability to adapt and navigate different requirements has been integral to our survival. It has helped us to inhabit many different landscapes, nurture and harness different food sources, live together in significantly larger groups and demonstrate our creativity through a range of entertainments along the way too. This truly is part of what makes us human.

Therefore, in spite of all of the above differences and changes we can expect to encounter within our brains and those of our audiences, I find that at some point practicality has to kick in. So, for me, I use the following hierarchy as a guideline to start from. On balance, these are the 'generally accepted' order in which our sensory information *could* be considered to operate.

1. sight
2. hearing

3 touch

4 smell

5 taste

I am not going to add all the caveats here because I feel you have already heard enough about those. Just please know that this is far from concrete but is instead offered as a pragmatic approach to enable you to make a start.

Examples of combined senses

Before we close this exploration, I thought it might be interesting, and in places amusing, to consider some examples of instances where combinations of our senses have been used in novel ways. Take for example the concept of 'dinner in the dark'.

For anyone who has not encountered this concept before, it fundamentally does what it sounds like. It is a series of restaurants across the world that serve you your meal in total pitch darkness. In some restaurants this is achieved by blindfolding their customers on arrival (so the staff are still able to operate in the low-light environment). However, other restaurants take it to the extreme of creating a total black-out environment (requiring the serving staff to be visually impaired or wear night-vision goggles to navigate their way around). The verdict appears to be very clear though – when you cannot see what you are eating and drinking, you focus your attention on the flavours and textures to a far greater extent. Even the sounds around you are reported to be more intense, as are the tactile experiences such as locating your glass and cutlery.

Instead of just depriving our brain of one sense in order to optimize the others, some concepts are based around trying to remove almost all of our incoming sensory information. Sensory deprivation tanks allow the body to float in a highly buoyant liquid that is the same temperature as the body. Lights and sounds are severely restricted, and smells are removed to create an almost total sensory absence. Short term, this can be highly relaxing. It is thought to relieve anxiety, improve sleep quality, induce calm, reduce stress and even improve symptoms of pain. But this is all short term. Depriving someone of their sensory information for an extended duration is torture. Literally. If used for a long period of time, the brain will start to create substitutes, leading to hallucinations, cognitive impairments, depression and severe anxiety.

On a lighter note again, the next time you are shopping, take a moment to notice the music that is being played in the background. In some instances, this may be influencing the decisions you make because the connections between your senses are working against you. For example, a study was conducted in a supermarket to explore the effect of playing different music on wine sales. For a period of two weeks, typically French and German music were played on alternating days. When they examined the sales figures for the same period, they found that on the days the French music was played, the sales of French wines exceeded sales of German wines by a ratio of 5:1. Conversely, on days when German music was played, sales of German wines exceeded sales of French ones by a ratio of 2:1. However, when customers completed a survey as part of the study, it became apparent that they were unaware of the way the background music influenced their product selections. This is yet more evidence of how we are influenced by elements that remain below our conscious threshold of awareness.

We are learning that our senses are far more connected than we had thought. The information they provide is often used to corroborate what is experienced through the other senses, and to create predictions of what we will experience. Although these processes are hard for us to manage or influence in some situations, it is important that we are aware of them.

Moreover, the interplay between our senses provides great justifications for us to develop campaigns and content that appeal to a number of our senses, as we know these immersive experiences are more likely to get noticed and attended to. For example, using sound, smell or touch to back up the visual aspects of a campaign can dramatically alter the results it will achieve. Where could you be adding more senses to the materials you produce?

Chapter summary

1. Although we have examined and explored each of our senses in isolation, in reality, they do not operate that way. Instead, they combine to provide verification, predictions and occasionally conflicts.
2. Despite numerous attempts to ascertain what it is, it appears that there is no one, single method that our brains use to prioritize the information that comes in from different senses.
3. Expectations created in the form of predictions can drastically change the way we interpret and make sense of incoming sensory information. This concept of priming can be used to shape the experience someone has and their feelings towards it.
4. The priorities allocated to incoming sensory information can vary dependent on, among other aspects, the context, previous experience, relevance and emotional state of the individual.
5. We also need to acknowledge that many cultural sensitivities can come into play when our sensory information combines, as our ability to articulate and describe our experiences actually shapes our sensation of them.
6. In practical terms, for most of us marketers operating in most business environments, we should consider prioritizing vision over the other senses, followed by hearing and touch in that order.
7. Some products, services and businesses are based around the interplay between our sensory information, and they can in turn produce changes in our physiology, experiences and behaviours as a result of emphasizing or removing the original inputs.
8. To be effective, most marketing campaigns should really look to incorporate more than one sense, allowing for more interactive, stimulating and memorable strategies, campaigns and communications.

Notes

1 Mitchell, M B, Workman, A D, Rathi, V K and Bhattacharyya, N (2023) Smell and taste loss associated with Covid-19 infection, *The Laryngoscope*, 133 (9), 2357–61, https://doi.org/10.1002/lary.30802 (archived at https://perma.cc/PE37-AJCE)

2 Herz, R S and Bajec, M R (2022) Your money or your sense of smell? A comparative analysis of the sensory and psychological value of olfaction, *Brain Sciences*, 12(3), DOI: 10.3390/brainsci12030299 (archived at https://perma.cc/4VNR-EGN9)

3 Brochet, F and Dubourdieu, D (2001) Wine descriptive language supports cognitive Specificity of chemical senses, *Brain and Language*, 77 (2), 187–96, https://doi.org/10.1006/brln.2000.2428 (archived at https://perma.cc/VW5A-YES4)

4 Khalaf, A, Lopez, E, Li, J, Horn, A, Edlow, B L and Blumenfeld, H (2025) Shared subcortical arousal systems across sensory modalities during transient modulation of attention, *NeuroImage*, 312, https://doi.org/10.1016/j.neuroimage.2025.121224 (archived at https://perma.cc/S57D-TPCH)

5 Majid A et al (2018) Differential coding of perception in the world's languages, *Proceedings of the National Academy of Sciences of the United States of America*, 115 (45), 11369–76, https://doi.org/10.1073/pnas.1720419115 (archived at https://perma.cc/3JPU-ZVMW)

15

Putting it all together

Context

Well done you! I am so proud of you for making it all the way through to this end chapter. You have learned so much about neuromarketing over the course of these pages, and here you still are, enthusiastic and curious to the end. I sincerely hope that you have not only found some insights that will help you to reach and connect with your audiences more effectively, but have also learned some interesting things about yourself too!

So, what happens now? Where do you go from here? And what support can you access as you begin to implement the concepts and content you have discovered?

The content in this book is designed to be an introduction to some of the vast array of content that has been discovered within the emerging field of neuromarketing. It is not, sadly, exhaustive. If we were to try and deliver that for you, it would be a never-ending task, as each week new discoveries are made, documented and debated within articles and journals across the world. However, I hope as a result of reading this book, two things have happened:

1. I hope you now feel more aware, informed and confident about neuromarketing, and your ability to utilize some of its major elements.
2. I also hope you are curious to continue your journey of discovery in this field, and will go forwards to uncover more insights and opportunities that will benefit your audiences and your organization.

I fully appreciate that some of the aspects I have introduced you to, and some of the tips and techniques we have discussed, are going to feel very different. Different from what you have done before, and potentially also very different from what you are doing now. That is OK. Much of what we have learned in marketing needs to be revised and updated as a result of insights from behavioural science and neuroscience. However, sadly, we have to accept that much of that is not going to change overnight. For now, I think the main message for you to take away is to just try not to feel too overwhelmed by what is ahead of you.

Most of us are not in the position to be able to totally overhaul all of our marketing materials and strategies as a result of learning these new insights. Instead, what typically happens is much more gradual. It is evolution rather than revolution, if you like. So, you may need to accept that you are going to have to live with some of your existing collateral and approaches for a while still, before you are able to make the sweeping changes that you may now desire. I would just say two things about this though. First, draw up a list of priorities. What is the most urgent item for you to attend to? By that I mean, what is most broken, most critical or most immediate for you to work on? Second, once you have identified where you want to begin, you may still need to be realistic about what you can change at this stage. It is not advisable to suddenly produce content or materials that differ wildly from what has gone before. Doing this may make your loyal customer base question their perceptions and understanding of you, and we do not want to do that. Again, think evolution...

So, with that thought firmly in the front of our minds, where should you start?

Well, for me the first step is... start! Honestly, just begin. Spoiler alert: what you produce in these early days will not be perfect. But I am 100 per cent confident they will still be an improvement on what you were previously producing before you had any neuromarketing insights to apply. That said, there are still elements that we can look at and utilize to improve your chances of getting some great results right from the outset.

1. Activities

In the resources that accompany this book, I have given you access to some tasks and activities that will give you some pointers relating to each chapter. My question to you then is have you done them? Any of them? Just some of them? None of them? If not, go to www.katiehart.co.uk/bookbonuses now and work your way through those. I have not created and collated those for my own benefit you know! These are designed specifically to help you be able to implement what you have learned, and also to get you to become more aware of examples around you from other providers and sectors.

In the process of doing these, you will build up a knowledge bank that you can refer to if you need a refresher or even just some inspiration going forwards. More than that though, while you are building your unique set of resources, I highly suggest you start a 'Library of Lessons' or an 'Exhibition of Examples'. This is something I have done for years, and continue to add to today. Having a large supply of examples (both good and bad!) to draw on and refer to boosts my creativity and gives me plenty of material for training courses and blogs. I call mine my 'Savings Bank' as usually it is me saving examples, but in all honesty, on some occasions, they have saved me! So, start yours today. It doesn't have to be attractive, clever or particularly well structured. Just create a repository where you can file electronic and hard copy examples that you encounter. For each one, I find it helpful to note:

1. **Date** To ensure I know when I came across it, and therefore how 'old' previous examples are. This also helps me to note any timely relevance, e.g. to a national occasion or newsworthy event.
2. **Context** Where did I encounter it? This could involve detailing the location, event, publication, platform, etc.
3. **Noteworthy** What is it about this example that stands out to me? Did it capture my attention in a crowded exhibition hall, make me think as I was listening and driving or draw me in through the visual design techniques adopted?
4. **Consideration** When I reflect on this through neuromarketing insights, what do I now also see? These may not necessarily be the reason I originally noticed it, but now that I consider it in more detail, what else do I appreciate?

5 **Clients** Here I note if this has triggered any specific thoughts or ideas for any of my existing or previous clients, so I am able to share this with them on the next project or campaign. This may not be relevant for you if you work in-house but for agencies it can be a great, and proactive, way to add value to your clients. Please note, I am not suggesting that you encourage your clients to copy what has already been done in their sector; I am referring to ideas taken out of context, from different industries and sectors, which just trigger off a thought or idea in you.

6 **Record** You will soon find this library builds into a large number of items, and that is to be celebrated. However, a corollary of this is that it does make it harder to search through when an example is sought. To that end, I have created an index for mine which allows me to easily draw on a wide selection of resources, ensuring they are relevant, timely and informative each time they are used.

Once you start to use these techniques and become more familiar with their application, you will soon feel much more confident in adopting them yourself. However, that will only come as a result of building your experience and exposure to the concepts you have learned. So, do some groundwork, get building your resources up, and just keep noticing what the organizations and brands you encounter are creating and using.

2. Secondary research

Now that you have got some basic knowledge under your belt, please do go out there and keep exploring neuromarketing. There are a number of scientific journals that feature and discuss neuromarketing discoveries, websites, podcasts, training courses, etc. The industry knowledge bank is increasing by the day, so get in amongst it and make sure you are able to continue your learning and build on the progress you have made so far.

I hope you recall that back in Chapter 2 I stated that one of my intentions for you as a result of reading this book was that you would be able to be discerning about the findings you uncover. You now know some of the methods used in neuromarketing, you understand the opportunities and challenges that they bring with them and all of this will stand you in good stead as you move forwards. So please do keep curious, keep exploring, and know that you do so from a position of insight and understanding now, so you will be able to determine much more effectively the relevance and likely impact of what you encounter.

3. Checklist

When I first started trying to implement some of the early discoveries that were made, I found that I needed some form of resource to support me. This was to ensure I remained within the realm of what neuroscience was teaching me, and not letting myself 'slip' back into the world of conscious rationalizations and thoughts. This was particularly vital when I was working on a number of projects simultaneously, for clients in sometimes very different sectors. In these instances, as any agency employee or owner will tell you, it is vital to be able to quickly immerse yourself in whichever account you are working on at that time. So, I produced my own checklist to ensure I was able to efficiently move from one project to another whenever the phone rang or an email came in!

I offer you a version of that checklist now. Please feel free to adapt or amend it as you need to, to optimize its effectiveness for your situation. Although I am not at all precious about how you structure this document, and what it looks like, I do care about one thing. Please do use something to support you. Much of what we have explored together is counter intuitive, and to keep yourself on track, it really does help to have some reference material that you can return to at key stages of the marketing process.

My checklist looks like Figure 15.1.

FIGURE 15.1 A suggested checklist for you to use as you begin to implement these concepts

Opportunity	
Audience	
Fears	
Flip it	
Proof	
Senses	
Emotions	
Attention	
Recall	
Decision-making	
Objections	
Close	
Call to action (CTA)	
Resources	

Let me talk you through what we have here.

1 **Opportunity** First, just identify which project, product or campaign this relates to. That is so you can compile and collate all the necessary information in one place, which pertains to the same piece of work. All of which is much more effective if you know which piece of work it refers to! So this may identify an individual campaign, presentation, piece of content or even something as big as a launch, research activity or a re-brand.

2 **Audience** State who the audience for this work will be. This should go into as much detail as you can, in terms of their demographics and basic persona. What do you already know about them? Now think about them through the neuromarketing approach, does that change anything for you? Or maybe there are things you still need to find out.
3 **Fears** Next, you need to try and identify what you feel their basic fears are. Drill right down to those unconscious, reptilian ones you uncovered, and note whether they are security, status or strength.
4 **Flip it** Now, what can you do to turn these around? How can you position your key messages as a solution to these underlying fears? This may feed into a wider picture that you are encouraging them to believe in, which addresses different fears you have identified.
5 **Proof** What evidence do you have to back up the claims you are making? Remember, they need to be addressed in the order of social, observable, analytical and lastly aspirational.
6 **Senses** Starting with the platforms you are going to be using, think about how you can play to their senses. Will it be in the language you use or can you actually incorporate different elements to harness more of their sensual experience?
7 **Emotions** What emotions are you likely to encounter? What is the likely state of your audience, and how can you work with that to support them? What emotions do you want to convey in your content, and how will you bring them to the fore? What considerations do we need to have regarding their emotions – is there one key aspect we need to address before they are open to being communicated with and helped?
8 **Attention** Now you have all of that scoped out, what devices are you intending to do to capture their attention. Can you use images, video, audio, etc? How will you include personalization, contrast, etc?
9 **Recall** How will you make the content more memorable? Make sure you include consideration of the primacy and recency effect, where you focus most on the content at the start and end of the piece/script/presentation/pitch. Do you have stories you can include? If so, how can you incorporate emotions and senses to bring the story to life and make it more likely to be recalled?

10 **Decision-making** How can you help their brain to see the easy comparisons and support their decision-making criteria? Make things as tangible as you can, and remember to keep the number of options small at each stage.

11 **Objections** What barriers can you anticipate that will prevent people from saying 'yes'? How can you support them with resolving these, and, again, remember to use the different forms of evidence to support your claims.

12 **Close** In addition to considering the primacy and recency effect, how do you want to end the piece? What do you want them to do, feel, change or notice? Can you apply the law of consistency and allow them to experience a small-scale version?

13 **Call to action (CTA)** What is the next step for them to take? How can you make this exciting, appealing, stand out, urgent, exclusive, etc? Remember to use what you have learned about colours, language and the numerous devices to attract attention and encourage people to act.

14 **Resources** What will you need in order to create this? If you are doing a presentation, can you use props or physical cues? If you are writing content for a landing page do you need up-to-date images or more recent testimonials? If you are recording video, which members of the team need to be involved and what locations will you use? Note them all down at this stage.

4. Split testing

One of the best ways for you to begin to work with these concepts is to try them out on your audience. Using technology, it has never been easier to create different versions of content and trial them. This could mean you segment off some of your audience and create a pilot for testing the materials out or it could mean you develop two or three concepts and randomly display them. Either way, the key here is to be able to accurately measure the results you achieve. And in order to do that, you need to just change *one* thing each time. Trust me, I know how exciting this content is and how enthusiastic you probably are by now to put it into action. However, if you create very different versions of content, you will never truly understand what is working.

For example, you might want to create two social media posts to test some of your ideas out. They might be designed as shown in Figure 15.2.

FIGURE 15.2 Two designs showing the dangers of changing too many variables between tests

Design 1	Design 2
• Image: A hi-tech manufacturing environment • Headline: Are you keeping up? • Text: Designed to show you understand their needs • Offer: 2 hr online workshop for £199 + VAT • CTA: Link to 'Sign up'	• Image: Someone lying awake at night • Headline: What's keeping you up? • Text: Designed to be 'real' and draw on their emotions • Offer: 2 hr in-person training for £499 + VAT • CTA: Link to 'Save my seat'

Now, if you ran both of these, randomly, for a period of a week, what conclusions could you draw? Say Design 1 achieved an 4.8 per cent conversion rate and Design 2 achieved a 6.1 per cent conversation rate. In what way are you really any the wiser? As frustrating as it may be, when we do research or tests like this, we need to only change one variable at a time. If we change more than that, we cannot be sure which of the things we changed was the cause of the different outcome. It may be that the headline for Design 2 was really effective, but people don't want to attend in-person courses. Or maybe the text on the CTA button was what achieved the increase in conversions. You will never know unless you isolate the variable and only change one thing at a time. Yes, it may be slow, but that way you are making definitive progress in the right direction with every test you put out.

Concluding remarks

And there you have it – your introduction to neuromarketing. I hope you have got lots of insights and ideas from this content, and I hope you remain curious and excited for what it can help you to achieve. Championing the human element within our profession is something I will never tire of doing, and I hope you share that passion now too.

Please know that you *can* do this, you *can* bring about significant changes, you *can* learn to recognize and implement many of the techniques I have introduced you to here. Underneath all the terminology, processes, analysis and overthinking, you are human too. Remember that. Celebrate that. Trust that.

It is what we all want from the brands and organizations we align ourselves with. Those brands who see the opportunities and accept the need to change have so much to gain.

I sincerely hope that you are now among that number.

INDEX

The index is filed in alphabetical, word-by-word order. Acronyms and 'Mc' are filed as presented; numbers are filed as spelt out in full. Locators in italics denote information within figures.

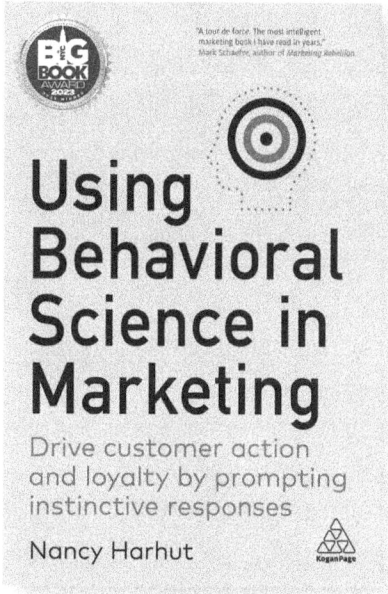

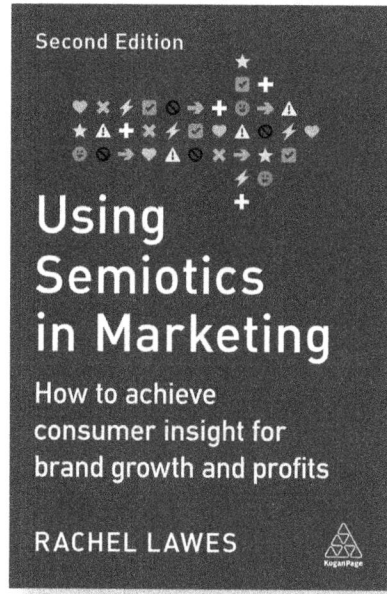

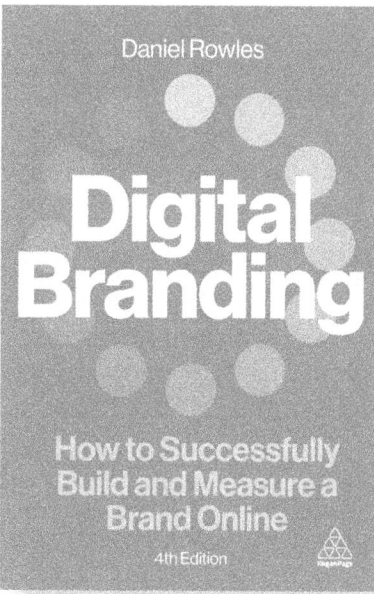

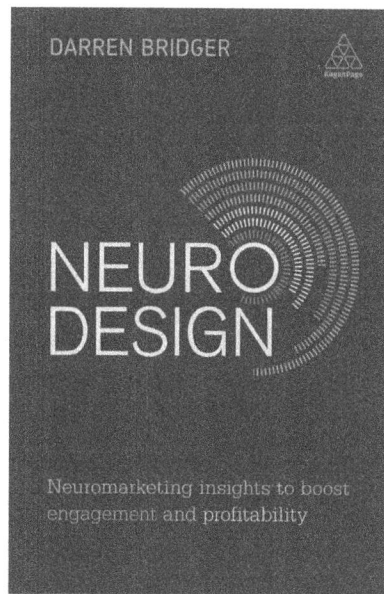

From 4 December 2025 the EU Responsible Person (GPSR) is:
eucomply oÜ, Pärnu mnt. 139b – 14, 11317 Tallinn, Estonia
www.eucompliancepartner.com

www.ingramcontent.com/pod-product-compliance
Lightning Source LLC
Chambersburg PA
CBHW071551210326
41597CB00019B/3195